Cambridge Tracts in Mathematics
and Mathematical Physics

No. 25

THE SYMMETRICAL OPTICAL SYSTEM

THE SYMMETRICAL OPTICAL SYSTEM

BY

G. C. STEWARD, Sc.D.

Professor of Mathematics in the University of Hull
Sometime Fellow of Gonville and Caius College, Cambridge

CAMBRIDGE
AT THE UNIVERSITY PRESS
1958

CAMBRIDGE
UNIVERSITY PRESS

University Printing House, Cambridge CB2 8BS, United Kingdom

Cambridge University Press is part of the University of Cambridge.

It furthers the University's mission by disseminating knowledge in the pursuit of education, learning and research at the highest international levels of excellence.

www.cambridge.org
Information on this title: www.cambridge.org/9781107493889

First printed 1928
Reprinted 1958
First published 1958
Re-issued 2015

A catalogue record for this publication is available from the British Library

ISBN 978-1-107-49388-9 Paperback

PREFACE

In the following tract is given an outline of the theory of the Symmetrical Optical System, both from the purely geometrical and also from the physical point of view. The first few chapters are based upon a course of lectures delivered to students of mathematics and of physics and the interest, to this extent, is theoretical; but the methods employed lend themselves readily to the computation of actual optical systems, and in Chapter V are given formulae used in optical calculations: it is hoped that these will be of practical interest. The treatment is based upon the Characteristic Function of Hamilton or else upon one of its modifications, the Eikonal; for, in my opinion, such a function offers by far the most powerful method of examining the behaviour of optical systems, whether from the theoretical or from the practical point of view. The geometrical meaning of the various aberrations is considered, both those of the first order and also those of higher orders, together with the more important of the Optical Conditions such, for example, as the recent Optical Cosine Law, of which the well-known Sine-Condition is but a particular case. And, inasmuch as the effect of the geometrical aberrations is very largely masked by diffraction phenomena, an account is given of the diffraction patterns associated with the optical system and the modifications of them due to these geometrical aberrations; moreover in addition to the usual circular aperture other forms of aperture also are considered, namely, the annular aperture, the slit aperture and the semi-circular aperture.

I am greatly indebted to Mr T. Smith, of the Optical Department of the National Physical Laboratory, for his kindness in reading the proofs of the tract; and I should like to record here my gratitude to him for his kindly encouragement in optical work and for many pleasant hours spent at the Laboratory.

I have also to render my acknowledgments to the Council of the Royal Society for their courteous permission to reproduce several diffraction diagrams, appearing in the last two chapters, which were published in a Paper communicated to the Society. And, finally, it remains for me to exprcss my thanks to the University Press for the usual and very high standard of their work.

G. C. S.

CAMBRIDGE
December, 1927

CONTENTS

CHAPTER I

ELEMENTARY THEORY

1. *Pure Geometry.* In the absence of aberrations a symmetrical optical system may be regarded as a means of transformation between two three-dimensional regions and the transformation is such that there is a one-to-one correspondence between points, lines and planes respectively of the regions. Moreover to a straight line bisected normally by the axis of the system will correspond another straight line bisected normally by the axis, owing to the symmetry of the system.

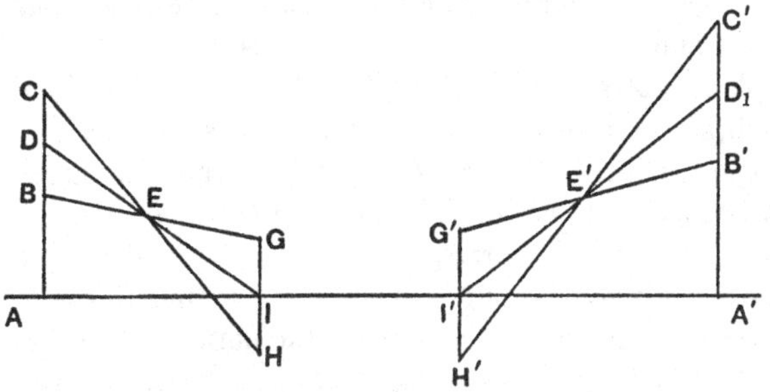

Let BC be a line normal to the axis AA' of a symmetrical optical system and let D be its mid-point; let $B'C'$ correspond to BC and let E and E' be corresponding points. Join DE intersecting AA' in I and let BE and CE intersect a normal to the axis through I in G and H; let $H'I'G'$ correspond to HIG and join $G'E'$ and $H'E'$ passing through B' and C', since G', E', B' and H', E', C' correspond to collinear points. Let $I'E'$ intersect $B'C'$ in D_1; then D_1 corresponds to D. Since $BD=DC$, then $HI=IG$ and $H'I'=I'G'$, and therefore $B'D_1=D_1C'$; thus D_1 is the mid-point of $B'C'$. The same property may be proved for oblique lines lying in a plane normal to AA'; thus to any straight line BDC lying in a plane normal to the axis of the system and bisected at D corresponds a straight line $B'D'C'$ also lying in a normal plane and bisected at D'.

Owing to the one-to-one correspondence in points there will be two homographic ranges lying upon the axis AA' of the system; to the 'point at infinity' in the direction $A'A$ will correspond a 'flucht' point F_2—which as a special case may itself be infinitely distant; and to the 'plane at infinity' in the direction $A'A$ will correspond a normal plane through F_2. Similarly a point F_1 and a normal plane through F_1 will correspond to the 'point at infinity' and the 'plane at infinity' in the direction AA'.

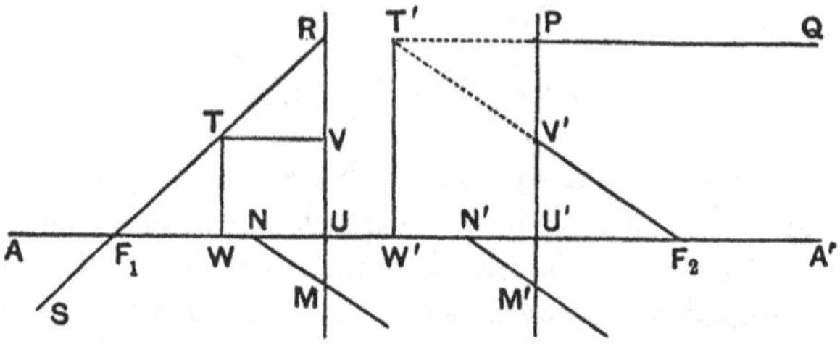

Let SF_1R and RPQ be corresponding lines; then RPQ will be parallel to AA' since it must meet AA' 'at infinity'; and since to a normal plane corresponds a normal plane the locus of R will be a plane normal to the axis and intersecting it in U. Corresponding to this will be a normal plane intersecting AA' in U' and it is clear that these same two planes will be obtained by proceeding from the second region to the first region. The four points F_1, F_2, U and U' determine the transformation completely and it is to be noticed that while U and U' are corresponding points F_1 and F_2 do not correspond. Let T be any point which without loss in generality may be taken upon F_1R; to find the corresponding point draw TV parallel to AA' meeting the normal plane through U in V and let V' be the corresponding point upon the normal plane through U'. Let $V'F_2$ intersect the line PQ in T'; then T' will correspond to T. Draw TW and $T'W'$ perpendicular to AA'; then W and W' will be corresponding points. Moreover

$$F_1W \,.\, W'T' = F_1U \,.\, WT \quad \text{and} \quad W'F_2 \,.\, WT = U'F_2 \,.\, W'T',$$

so that

$$F_1W \,.\, F_2W' = -F_1U \,.\, U'F_2.$$

Let N and N' be two points upon the axis AA' such that $F_1N = U'F_2$ and $F_1U = N'F_2$; then N and N' are corresponding points. Let any line through N intersect the normal plane through U in M and let $N'M'$ be the corresponding line intersecting the normal plane through U' in M'. Then $UM = U'M'$ and $NU = N'U'$, so that NM and $N'M'$ are parallel. Thus corresponding to any straight line through N there will be a straight line through N' and these two straight lines will be parallel.

2. The transformation considered in the previous paragraph is uniquely determined when F_1 and F_2 are given in position and also one of the pairs of points U, U' and N, N'; these six points are named therefore 'cardinal points' and the normal planes through them 'cardinal planes.'

F_1 and F_2 are called 'principal foci,' U and U' 'unit points' and N and N' 'nodal' points. The following relations hold, therefore:

$$F_1U = N'F_2 = f \quad \text{and} \quad F_1N = U'F_2 = f';$$

and f and f' are named respectively the 'first' and 'second focal lengths.' No relation between f and f' can be obtained from pure geometry and recourse must be had to a law of optics which states that 'the ratio of f to f' depends only upon the optical properties of the two regions in which they are measured': this will be proved subsequently*.

In the figure of §1 let us write $F_1W = x$, $F_2W' = x'$ and $W'T' = mWT$, so that m is the 'magnification' or the ratio of the 'image' $W'T'$ to the object WT; then from the relations given there we have

$$x = \frac{1}{m}f \quad \text{and} \quad x' = -mf',$$

so that
$$xx' = -ff'.$$

This result is due to Newton and it gives the point upon the axis in either region corresponding to a given point upon the axis in the other region and also the magnification associated with any pair of conjugate points; it is seen that the origins of co-ordinates are the principal foci and that the directions of measurement are the same.

3. Before considering formulae applicable to particular systems it may be as well to obtain a few more general results. Thus let F_1 and F_2 be the principal foci and U and U' the unit points of a symmetrical optical system

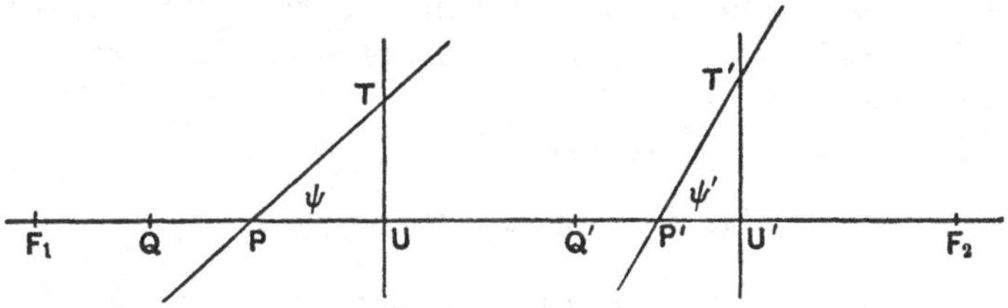

and let P, P' and Q, Q' be pairs of corresponding axial points associated respectively with magnifications m and n; and let x, x' and y, y' be the co-ordinates of these points referred to the principal foci as origins; we intend to change these origins to the conjugate points Q and Q', referred to which let the co-ordinates of P and P' be u and v respectively.

Then
$$u = QP = F_1P - F_1Q = f(1/m - 1/n)$$
and
$$v = Q'P' = F_2P' - F_2Q' = -f'(m - n)$$

* Cf. chapter II, § 3.

from the preceding paragraph; so that

$$nf'/v - f/nu = 1. \quad \text{.........................(1)}$$

If Q and Q' coincide with the unit points, $n = 1$, and then

$$f'/v - f/u = 1; \quad \text{.........................(2)}$$

if f and f' be equal we have

$$1/v - 1/u = 1/f, \quad \text{.........................(3)}$$

while if f and f' become infinite while their ratio remains finite according to the relation

$$f/\mu = f'/\mu',$$

μ and μ' being, so far, undefined constants, we have

$$\mu'/v - \mu/u = 0. \quad \text{.........................(4)}$$

This relation is of use in 'telescopic' systems.

Again, in the general case, we may define, with Maxwell, the ratio of $P'Q'$ to PQ to be the 'elongation'; and we have

$$\text{elongation} = \frac{P'Q'}{PQ} = \frac{F_2Q' - F_2P'}{F_1Q - F_1P} = \frac{(m-n)f'}{(1/n - 1/m)f} = mnf'/f. \ldots(5)$$

The longitudinal magnification at any point upon the axis is proportional therefore to the square of the transverse magnification associated with the point; and we may find similarly the oblique magnification corresponding to any oblique angle.

4. Again let PT be any line through P, intersecting the first unit plane in T and cutting the axis at an angle ψ; let $P'T'$ be the corresponding line intersecting the second unit plane at T' and cutting the axis at an angle ψ'. Then $UT = U'T'$ and

$$\therefore\ PU \tan\psi = P'U' \tan\psi',$$

i.e.,

$$f\left(1 - \frac{1}{m}\right)\tan\psi = -f'(1-m)\tan\psi',$$

$$\therefore\ f\tan\psi = mf'\tan\psi',$$

i.e.,

$$\tan\psi/\tan\psi' = mf'/f^{*}. \quad \text{.......................(1)}$$

If l and l' be the lengths of two corresponding normal lines at P and P' we have $l' = ml$ and then (2) takes the form

$$lf\tan\psi = l'f'\tan\psi'\dagger, \quad \text{.......................(2)}$$

* This ratio is named, by Southall, the 'Angular Magnification' or 'Convergence-Ratio.'

† Attributed to various writers, e.g., Helmholtz, Lagrange; but it appears in Robert Smith's *Compleat Opticks*, Cambridge, 1738, for a system of thin lenses.

i.e., this quantity is unaltered by the transformation. If the angles involved be small we have

$$lf\psi = l'f'\psi'$$

or
$$\mu l\psi = \mu' l' \psi' \quad \text{.........................(3)}$$

if we assume that $f/\mu = f'/\mu'$.

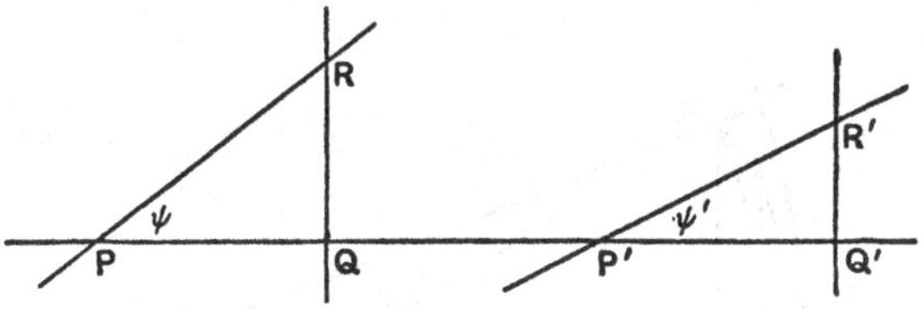

Let now the line through P intersect the normal plane through Q in R where $QR = y$; and let the corresponding line through P' intersect the normal plane through Q' in R' where $Q'R' = y'$. Then

$$Q'R' = ny = P'Q' \tan\psi'$$
$$= f'(m - n)\tan\psi'$$
$$= f\tan\psi - nf'\tan\psi' \quad \text{from (1)}$$

and this is a constant quantity for rays through R'.

This may be written

$$yy' = fy\tan\psi - f'y'\tan\psi', \quad \text{.....................(4)}$$

a symmetrical relation. It may be shewn that for lines not meeting the axis of the system their projections upon a plane passing through the axis will satisfy this relation.

The preceding results have all been obtained from a purely geometrical theory of 'collineation'—shewing how much may be derived from the mere notion of 'images'; no optical principle has been introduced so far. We shall see later that the condition that rays from a point near to P and upon the normal plane through P should be brought accurately to a focus at a point near to P' and upon the normal plane through P' is

$$f\sin\psi = mf'\sin\psi', \quad \text{.....................(5)}$$

and this will be derived from optical principles. This is clearly inconsistent with equation (2) and we draw the conclusion that a 'perfect' optical system is an impossibility*.

5. ***Refraction at a Single Spherical Surface.*** The preceding paragraphs have dealt with a purely geometrical transformation and we have now to make an optical application. It will be assumed for the purposes of this

* Cf. Maxwell, 'On the General Laws of Optical Instruments,' *Sci. Papers*, vol. I, pp. 271–285; Southall, *Geometrical Optics*, chap. VIII.

tract that light is propagated under the form of a wave motion and that the disturbance originating at a point of a homogeneous medium is to be found at a subsequent time upon a concentric sphere—the radius of the sphere being proportional to the time interval and depending also upon the nature of the medium.

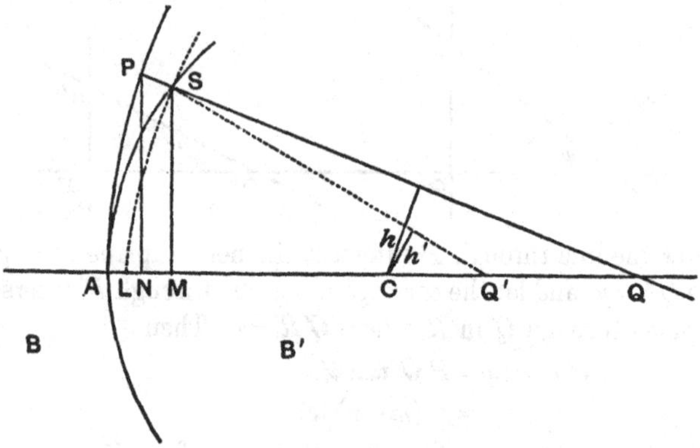

Let P be a point upon a spherical wave-front which touches at A the spherical boundary, centre C, between two media B and B'; let a normal at P to the wave surface cut AC produced in Q and let the angle PQA be small; let this normal cut the bounding surface in S and draw NP and MS normal to AC intersecting AC in N and M respectively. After time t the disturbance from P will have reached S and that from A will have reached L, upon AC; draw the sphere with centre on AC and passing through L and S. It will be shewn that to the first approximation this sphere is the new wave-front.

Let U and R be the curvatures of the incident wave-front and of the bounding surface respectively, so that

$$U = 1/AQ \quad \text{and} \quad R = 1/AC.$$

Then if $NP = y$ we have

$$2AN = y^2 U \quad \text{and} \quad 2AM = y^2 R;$$

this is an approximation and assumes that $NP = MS$. Now the distance travelled by the light from P to the sphere is, to the same degree of approximation, NM,

i.e., $$AM - AN = \tfrac{1}{2} y^2 (R - U).$$

If v be the velocity of the luminous disturbance in the medium B,

$$2vt = y^2 (R - U);$$

and similarly if v' be the velocity in the medium B' and V the curvature of the symmetrical sphere through L and S,

$$2v't = y^2(R-V).$$

Thus $$\frac{v}{v'} = \frac{R-U}{R-V} = \frac{\mu'}{\mu}\text{ (say)}, \quad\text{.......................(1)}$$

and this relation does not involve y; thus given μ, μ', R and U there is a unique value of V, independent of t and y, i.e., of the point chosen upon the original wave-front. The symmetrical sphere through L and S is therefore the new wave-front in the medium B'.

The quantities μU and μV may be named the 'equivalent' curvatures and (1) takes the form

$$\mu' V - \mu U = (\mu' - \mu) R \quad\text{.......................(2)}$$

or $$\Delta(\mu U) = (\mu' - \mu) R = \kappa\text{ (say)}, \quad\text{..................(3)}$$

where Δ is the usual operator of difference. Thus the effect of the refraction is to increase the equivalent curvature of the incident wave-front by the constant quantity $(\mu' - \mu) R$ or κ and this quantity depends only upon the properties of the media and the curvature of the bounding surface between them; it may therefore be named the 'power' of this surface.

Moreover $$\frac{\sin CSQ}{\sin CSQ'} = \frac{CQ}{SQ}\cdot\frac{SQ'}{CQ'} = \frac{R-U}{R-V} = \frac{\mu'}{\mu},$$

i.e., $$\mu' \sin CSQ' = \mu \sin CSQ,$$

to the degree of approximation contemplated.

6. If Q' be the centre of the new wave-front it will be seen that between the points Q and Q' there is a one-to-one correspondence; moreover the line of propagation PS of the incident disturbance has been transformed into the new line SQ' of propagation of the disturbance; we may employ therefore the geometrical transformation considered in the previous paragraphs. It will be observed that the results obtained are legitimate only if we neglect small quantities; the transformation given therefore is true only as a first approximation and in order to allow for this we must consider the 'aberrations' of the optical system—which accordingly will be effected in subsequent chapters.

A special case of §5 arises when the incident disturbance is reflected at the bounding surface AS—as indeed will always be the case partially. But here we may write $\mu' = -\mu$ and remember that the disturbance is to be regarded as travelling positively after incidence in the direction CA.

From (1) §5 we have $$\mu' v' = \mu v,$$

so that μ is a constant of the medium—varying indeed inversely as the velocity of the luminous disturbance in the medium; and it may be made definite by writing

$$\mu' v' = \mu v = v_0, \quad \text{.............................(1)}$$

where v_0 is the velocity of light *in vacuo*.

A particularly simple application of (3) §5 may be considered. Consider two spherical surfaces placed close together and touching at A and let the medium between them be defined by the constant μ—the outer media being the same and defined by the constant unity. Then a double application of (3) §5 and an obvious notation leads to

$$\mu V_1 = U_1 + \kappa_1 \text{ and } V_2 = \mu U_2 + \kappa_2;$$

where κ_1 and κ_2 are the powers of the surfaces. But $V_1 = U_2$ since the surfaces touch and therefore

$$V_2 = U_1 + \kappa_1 + \kappa_2.$$

Thus the effect of a 'thin lens' is merely to make a constant addition to the curvature of the incident wave-front.

7. ***Various Formulae.*** The formulae (2) and (3) § 5 are the fundamental formulae for refraction at a single spherical surface; we may notice some other forms of these however which will be of use subsequently. In the first place we may change the origin of co-ordinates from the pole A to the centre of curvature C; and it is easily verified that the result is

$$\frac{1}{\mu' CQ'} - \frac{1}{\mu CQ} = \frac{\mu' - \mu}{\mu\mu'} R$$

$$= \frac{\kappa}{\mu\mu'} = J. \quad \text{.....................(1)}$$

Here J, defined by the last relation, may be named the 'modified power' of the surface; further the expression μCQ will be called the 'reduced' distance CQ, i.e., it is the geometrical distance multiplied by the constant μ of the medium in which the distance is measured. The physical meaning of this product will be examined subsequently; (1) now becomes

$$\frac{1}{CQ'} - \frac{1}{CQ} = J, \quad \text{.........................(2)}$$

if we agree to regard all distances as reduced. Here the constant μ does not appear explicitly and this will be found to be true generally if the modified power be used instead of the ordinary power.

Again formulae involving angles are frequently of use and (2) §5 may be modified to this end. Thus if SQ and SQ' cut the axis AQQ' at angles α and α' respectively (fig. §5) we have

$$y = \alpha . AQ \text{ and } y = \alpha' . AQ' \text{ approximately.}$$

Substituting we have

$$\mu' a' - \mu a = \kappa y, \quad \text{.........................(3)}$$

i.e.,

$$\Delta(\mu a) = \kappa y. \quad \text{.............................(4)}$$

8. *The General System—the K-Formulae.* The symmetrical optical systemwill be composed, in general, of coaxial spherical surfaces separating various media; let us consider the optical properties of such a system.

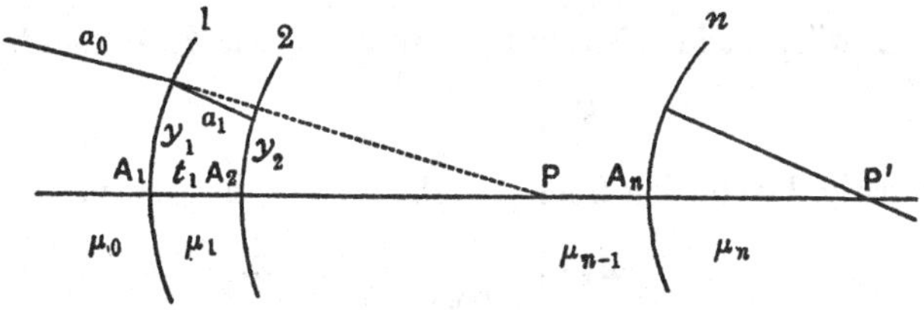

Let n coaxial spherical surfaces, 1, ... n, separate media whose optical constants are $\mu_0, \dots \mu_n$ and let $\kappa_1, \dots \kappa_n$ be the powers of the surfaces; let $A_1, \dots A_n$ be the poles of the surfaces, i.e., the points of intersection with the axis of the system. Let a ray inclined at an angle a_0 with the axis in the first medium meet the first surface in a point at distance y_1 from the axis; for the corresponding ray in the second medium let these quantities be a_1 and y_2 respectively; and so on. Let P be the intersection of the ray a_0 with the axis, P_1 that of the ray a_1, and so on; let the separations of the surfaces be t_1, t_2, ... t_{n-1} so that, e.g., $A_1A_2 = t_1$. Then we have the following relations;

$$y_1 = A_1P \,.\, a_0,$$
$$\mu_1 a_1 - \mu_0 a_0 = \kappa_1 y_1,$$
$$y_1 - y_2 = t_1 a_1,$$
$$\dots\dots\dots\dots$$
$$\mu_n a_n - \mu_{n-1} a_{n-1} = \kappa_n y_n;$$

which may be written

$$y_1 = \bar{u}\mu_0 a_0,$$
$$\mu_1 a_1 - \mu_0 a_0 = \kappa_1 y_1,$$
$$y_2 - y_1 = -a_1\mu_1 a_1,$$
$$\dots\dots\dots\dots$$
$$\mu_n a_n - \mu_{n-1} a_{n-1} = \kappa_n y_n,$$

where $\bar{u}\,(\equiv A_1P/\mu_0)$ is the 'equivalent' distance AP and $a_\lambda \mu_\lambda - t_\lambda = 0$. From this it is evident that $\mu_\lambda a_\lambda / \mu_0 a_0$ is the numerator of the 2λth convergent to the continued fraction

$$\bar{u} + \frac{1}{\kappa_1 -}\,\frac{1}{a_1 +}\,\frac{1}{\kappa_2 -} \dots \frac{1}{+\,\kappa_n}, \quad \text{....................(1)}$$

while $y_\lambda/\mu_0 a_0$ is the numerator of the $(2\lambda-1)$th convergent to this continued fraction. We may denote by $K_{1,n}$ the numerator to the last convergent to the continued fraction

$$\kappa_1 - \frac{1}{a_1 +}\frac{1}{\kappa_1 -}\frac{1}{a_2 + \dots +}\frac{1}{\kappa_n}.$$

Then we have the relation

$$\frac{\mu_n a_n}{\mu_0 a_0} = \overline{u}K_{1,n} + \frac{\partial K_{1,n}}{\partial \kappa_1}. \quad \text{.....................(2)}$$

If $a_n = 0$, P will coincide with the first principal focus F_1; so that from (2)

$$F_1 A_1 = \frac{\mu_0}{K_{1,n}}\frac{\partial K_{1,n}}{\partial \kappa_1}, \quad \text{.......................(3)}$$

and similarly, by starting from the other end of the system,

$$A_n F_2 = \frac{\mu_n}{K_{1,n}}\frac{\partial K_{1,n}}{\partial \kappa_n}. \quad \text{.......................(4)}$$

Again P will coincide with the first nodal point if $a_0 = a_n$; so that from (2)

$$\frac{\mu_n}{\mu_0} = \frac{A_1 N}{\mu_0} K_{1,n} + \frac{\partial K_{1,n}}{\partial \kappa_1};$$

whence $$F_1 N = \frac{\mu_n}{K_{1,n}} = f', \text{ the second focal length; ....(5)}$$

similarly the first focal length is given by

$$f = \frac{\mu_0}{K_{1,n}}. \quad \text{.............................(6)}$$

We verify therefore the general relation

$$f'/\mu_n = f/\mu_0 \text{*}. \quad \text{.............................(7)}$$

9. The quantities $K_{1,\lambda}$ and $\frac{\partial K_{1,\lambda}}{\partial \kappa_\lambda}$ are of importance from the point of view of subsequent aberration theory; and the following relations are given from which they can be calculated step by step. Let p_l be the numerator of the last convergent to the continued fraction

$$\kappa_1 - \frac{1}{a_1 +}\frac{1}{\kappa_2 -}\frac{1}{a_2 + \dots +}\frac{1}{\kappa_\lambda};$$

then $p_l = K_{1,\lambda}$, and

$$K_{1,\lambda} = \kappa_\lambda p_{l-1} + p_{l-2}, \text{ so that } p_{l-1} = \frac{\partial K_{1,\lambda}}{\partial \kappa_\lambda};$$

thus $$K_{1,\lambda} = K_{1,\lambda-1} + \kappa_\lambda \frac{\partial K_{1,\lambda}}{\partial \kappa_\lambda} \quad \text{.....................(1)}$$

and $$\frac{\partial K_{1,\lambda}}{\partial \kappa_\lambda} = \frac{\partial K_{1,\lambda-1}}{\partial \kappa_{\lambda-1}} - a_{\lambda-1}K_{1,\lambda-1}. \quad \text{..................(2)}$$

* Chap. II, § 3.

Similarly, starting from the other end of the system, we have

$$K_{\lambda,n}=K_{\lambda+1,n}+\kappa_\lambda\frac{\partial K_{\lambda,n}}{\partial\kappa_\lambda},\quad \frac{\partial K_{\lambda,n}}{\partial\kappa_\lambda}=\frac{\partial K_{\lambda+1,n}}{\partial\kappa_{\lambda+1}}-a_\lambda K_{\lambda+1,n}.\ \ldots(3)$$

From these three results the quantities $K_{1,\lambda}$ and $\frac{\partial K_{1,\lambda}}{\partial\kappa_\lambda}$, which are needed in optical computation, may be calculated readily step by step. Moreover from the above continued fraction it follows, by the elementary theory of continued fractions, that

$$\frac{\partial K_{1,n}}{\partial\kappa_1}\cdot\frac{\partial K_{1,n}}{\partial\kappa_n}-K_{1,n}\frac{\partial^2 K_{1,n}}{\partial\kappa_1\partial\kappa_n}\equiv 1,\ \ldots\ldots\ldots\ldots\ldots\ldots(4)$$

a result due to Gauss; and further, using the fraction of the preceding paragraph,

$$\overline{uv}K_{1,n}+\bar{u}\frac{\partial K_{1,n}}{\partial\kappa_n}+\bar{v}\frac{\partial K_{1,n}}{\partial\kappa_1}+\frac{\partial^2 K_{1,n}}{\partial\kappa_1\partial\kappa_n}=0,\ \ldots\ldots\ldots\ldots(5)$$

where $\bar{v}$ is the equivalent distance of the conjugate point from the last surface of the system.

Again from (1) §8

$$\frac{y_\lambda}{y_1}=\frac{\partial K_{1,\lambda}}{\partial\kappa_\lambda}+\frac{1}{\bar{u}}\frac{\partial^2 K_{1,\lambda}}{\partial\kappa_1\partial\kappa_\lambda}\rightarrow\frac{\partial K_{1,\lambda}}{\partial\kappa_\lambda},\ \ldots\ldots\ldots\ldots\ldots\ldots(6)$$

if $\bar{u}\rightarrow\infty$; thus a ray parallel to the axis and incident at a height y_1 upon the first surface will intersect the surface λ at a height given by (6).

From §6 it is evident that we may replace our refracting surfaces by thin lenses and the results of §8 will be similar in form; so that for example the combined power of two thin lenses κ_1 and κ_2 at a distance a apart is

$$K=\kappa_1+\kappa_2-a\kappa_1\kappa_2.$$

If therefore a exceed the combined focal lengths of the thin lenses, i.e., if the distance be positive between the second focus of the first lens and the first focus of the second, it follows that K will be negative, assuming κ_1 and κ_2 to be positive; a remark the bearing of which will be evident later.

10. *The J-Formulae.* From the definition in §7 of the modified power it is evident that corresponding to the preceding results involving the ordinary power κ there will be similar results involving the modified power J. In the figure of §5 let h and h' be the perpendicular distances from the centre C upon SQ and SQ'; then

$$h=CS\sin CSQ\quad\text{and}\quad h'=CS\sin CSQ';$$

so that $$\mu h=\mu' h'=\mathfrak{h}\ ,\ \ldots\ldots\ldots\ldots\ldots\ldots\ldots\ldots\ldots(1)$$

i.e., the quantity μh is unaltered by refraction. Moreover

$$CQ\,.\,\alpha=h\quad\text{and}\quad CQ'\,.\,\alpha'=h';$$

and also for two consecutive surfaces

$$\mathfrak{h}_1 - \mathfrak{h}_2 = a_1 \alpha_1, \quad \text{..........................(2)}$$

where a_1 is the reduced distance between their centres of curvature. Using therefore (2) § 7, and remembering that all distances are now reduced, we have for a coaxial system

$$\mathfrak{h}_1/\alpha_0 = C_1 P^*,$$
$$\alpha_1/\alpha_0 - 1 = J_1 \mathfrak{h}_1/\alpha_0,$$
$$\mathfrak{h}_2 - \mathfrak{h}_1 = -a_1 \alpha_1/\alpha_0,$$
$$\dots\dots\dots\dots$$
$$\alpha_n/\alpha_0 - \alpha_{n-1}/\alpha_0 = J_n \frac{\mathfrak{h}_n}{\alpha_0},$$

and these are *formally* the same as those of § 8. Thus we may define the modified power $J_{1,n}$ as the numerator of the last convergent to the continued fraction

$$J_1 - \frac{1}{a_1 +} \frac{1}{J_2 -} \dots + \frac{1}{J_n},$$

and we have the relations

$$J_{1,\lambda} = J_{1,\lambda-1} + J_\lambda \frac{\partial J_{1,\lambda}}{\partial J_\lambda}, \qquad \frac{\partial J_{1,\lambda}}{\partial J_\lambda} = \frac{\partial J_{1,\lambda-1}}{\partial J_{\lambda-1}} - a_{\lambda-1} J_{1,\lambda-1},$$

$$J_{\lambda,n} = J_{\lambda+1,n} + J_\lambda \frac{\partial J_{\lambda,n}}{\partial J_\lambda}, \qquad \frac{\partial J_{\lambda,n}}{\partial J_\lambda} = \frac{\partial J_{\lambda+1,n}}{\partial J_{\lambda+1}} - a_\lambda J_{\lambda+1,n}.$$

Moreover the (reduced) distances of the principal foci from the initial and final centres of curvature are given by

$$^*F_1 C_1 = \frac{1}{J_{1,n}} \frac{\partial J_{1,n}}{\partial J_1} \quad \text{and} \quad C_n F_2 = \frac{1}{J_{1,n}} \frac{\partial J_{1,n}}{\partial J_n};$$

and also $F_1 N = N' F_2 = 1/J_{1,n}$. Again for an incident ray parallel to the axis of the system it may be shewn that $\frac{\partial J_{1,\lambda}}{\partial J_\lambda}$ is proportional to the perpendicular upon the ray in medium λ from the centre of curvature of surface λ.

11. *The 'Modified' notation.* It will be seen from § 8 that if the system be 'thin,' i.e., $t_\lambda = 0$ for all values of λ, then $K_{1,n} = \sum_{\lambda=1}^{n} \kappa_\lambda$; while if the system be formed of concentric surfaces, i.e., $a_\lambda = 0$ for all values of λ, then $J_{1,n} = \sum_{\lambda=1}^{n} J_\lambda$. Thus the κ-formulae are adapted for thin systems and the J-formulae for concentric systems. But further, the μ-constants of the

* A reduced distance regarded as being in the medium μ_0.

media do not appear explicitly if J-formulae be used; thus Newton's formula § 2 may be written

$$xx' = -ff' = -\frac{1}{\mu_0 \mu_n J^2_{1,n}},$$

i.e.,

$$xx' = -\frac{1}{J^2_{1,n}},$$

if reduced values of x and x' be used; or we may write

$$XX' = -1, \qquad \ldots\ldots(1)$$

where $X (\equiv xJ_{1,n})$ and $X' (\equiv x'J_{1,n})$ are named 'modified reduced distances.' Also we have $X = \frac{1}{m}$ and $X' = -m$; but here m is the reduced magnification, i.e., the ratio of the reduced size of the image to the reduced size of the object. These properties of the J-formulae will be found of considerable use subsequently.

12. *Addition of Systems.* Two symmetrical optical systems upon the same axis may be regarded as one system and simple formulae may be found to give the combined power in terms of the separate powers.

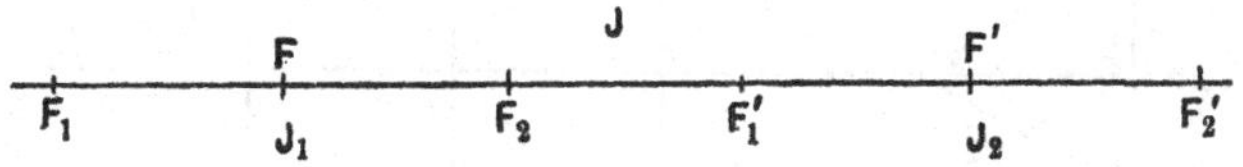

Let F_1 and F_2, F_1' and F_2' be the principal foci of the two coaxial systems of modified powers J_1 and J_2 respectively; let F and F' be the principal foci of the combined system and J its power. Then F and F_1' are corresponding points for the first system so that

$$F_1F \, . \, F_2F_1' = -\frac{1}{J_1^2},$$

all distances being reduced. Similarly considering the points F_2 and F' for the second system we have

$$F_1'F_2 \, . \, F_2'F' = -\frac{1}{J_2^2};$$

and from the points F_1 and F_2' and the combined system

$$FF_1 \, . \, F'F_2' = -\frac{1}{J^2}.$$

Whence

$$F_2F_1' = \pm J/J_1J_2, \quad F'F_2' = \pm J_1/J_2J \text{ and } F_1F = \pm J_2/JJ_1.$$

The ambiguity in sign may be cleared up by considering a special case; thus consider the addition of two thin lenses of positive power, the distance between them being greater than the sum of their focal lengths, so that F_2F_1' is positive here. The combined power is negative (§ 9), so that the negative sign must be chosen in the first result above, i.e., we have

$$F_2F_1' = -J/J_1J_2, \quad F'F_2' = +J_1/J_2J \text{ and } F_1F = +J_2/JJ_1. \quad \ldots(1)$$

13. *The Aperture Stop and the Field Stop.* The light passing through a symmetrical optical system is limited by the sizes of the refracting surfaces and also, in general, by diaphragms; which usually are circular with their centres upon the axis of the system. Let $A_1, A_2, \ldots$ be the refracting surfaces of a symmetrical system of which AA' is axis; let PQR be an object normal to AA'. Let S be any diaphragm or 'stop' of the system, including, as special cases, the rims of the refracting surfaces; then S divides the optical system into two parts—the one to the left of S and the other to the right of S. Let S' be the image of S in the part of the system to the left; we may treat each stop of the system in this

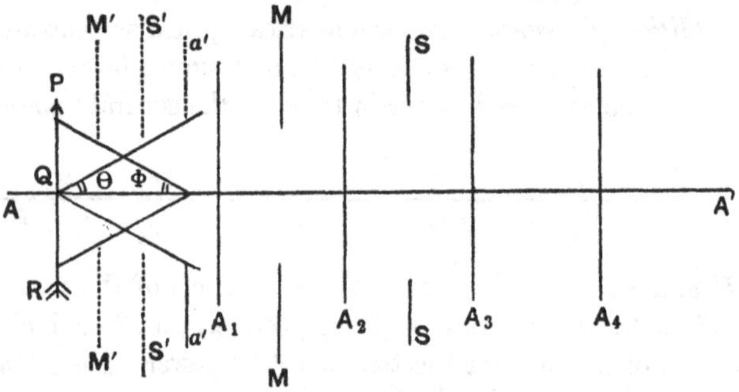

way and accordingly we shall have a number of images such as S', from which we may choose that one, α', which subtends the least angle, 2Θ, at the axial point of the object. This image α' is said to be the 'entrance-pupil' of the optical system, while the corresponding real stop α is named the 'aperture' stop; similarly there will be an image α'', given by the right-hand part of the system, and this image is known as the 'exit-pupil.' It is seen that Θ is the semi-vertical angle of the largest cone of rays, emanating from Q, which can pass through the system; accordingly Θ is known as the 'aperture angle.' Any ray leaving Q in a direction which makes an angle with the axis greater than Θ will be unable to pass through the system.

Again from the various images S', formed by the left-hand parts of the optical system, we may choose that one, M', which subtends the smallest angle, 2Φ, at the axial point of the entrance-pupil; and this image is said to be the 'entrance-window,' while the corresponding real stop, M, is named the 'field' stop. Similarly there will be an image, M'', on the right-hand side, known as the 'exit-window'; but of course all these names may be interchanged if we regard the light as travelling in the opposite direction. The least angle, Φ, defined by the entrance-window is known as the 'angular field of view' of the system; and it is clear that the function of the field stop is to limit the extent of the object, reproduced in the image by pencils, whose central ray passes through the axial point of the entrance-pupil; while the aperture stop limits the rays passing through the system. Moreover the precise stop which acts as field stop or as aperture stop of the system depends upon the position of the object.

If a small luminous object be placed, normal to the axis, at Q the energy passing from this object through the entrance-pupil, and, in consequence, through the system, is not proportional to the aperture angle Θ; for the intensity of radiation in any direction from Q will depend upon that direction. And, in fact, the energy passing through the system may be shewn to be proportional to the square of the quantity $\mu \sin \Theta$, μ being the optical index of the object space. This quantity $\mu \sin \Theta$ is known as the 'Numerical Aperture'* of the system, and plays a very important part in the theory of the microscope.

14. *Aplanatism.* Let there be a sphere of radius r and centre C of a medium of index μ where $\mu > 1$; let $CP'P$ be a diameter of the sphere and let $CP = r\mu$ and $CP' = r/\mu$.

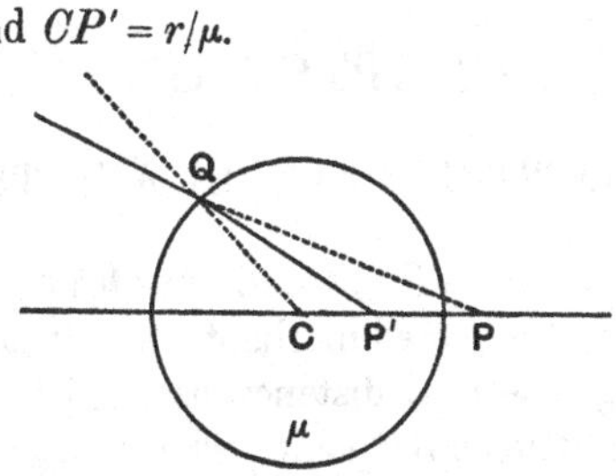

Then P and P' are inverse points with respect to the sphere and if Q be any point upon the surface

$$\frac{\sin CQP}{\sin CQP'} = \mu, \quad \text{.........................(1)}$$

and this without any approximation; so that *any* ray converging to P will, after falling upon the spherical surface, converge to P'. Moreover

$$\frac{\sin CPQ}{\sin CP'Q} = \frac{1}{\mu} = \text{const.} \quad \text{.....................(2)}$$

* See E. Abbe, 'On the Estimation of Aperture in the Microscope,' *Journ. Roy. Micr. Soc.* (2) I (1881), pp. 388–423.

Two points P and P' are said to be 'aplanatic' when both these conditions are satisfied, i.e., when rays incident towards either pass accurately after refraction through the other and when in addition the 'sine-condition' (2) is satisfied. The meaning of this latter condition will be discussed later; here it is sufficient to say that it ensures an accurate point image of a point object slightly removed laterally from P.

Now if the 'sine-condition' (2) is to be satisfied we must have

$$QP/QP' = \text{const.}, \qquad (3)$$

and from this it follows that the only single surface at which aplanatic refraction is possible is the sphere. We may indeed have an accurate point to point correspondence over any angle by using the surface of revolution generated by a Cartesian oval; but the sine-condition will not be satisfied and the refraction therefore is not aplanatic. Indeed considerable aberrations are introduced by refraction at such a surface.

REFERENCES

Czapski. *Die Theorie der Optischen Instrumente nach Abbe.*

Maxwell. On the General Laws of Optical Instruments. *Sci. Papers*, vol. I, pp. 271–285.

Southall, J. P. C. *Geometrical Optics*, chs. VI, VII, XIV.

Wandersleb. *Die Theorie der Optischen Instrumente nach Abbe* (ed. by M. von Rohr).

CHAPTER II

THE CHARACTERISTIC FUNCTION AND THE EIKONAL

1. *The Reduced Path and its Properties.* The idea of 'reduced' distance has been introduced in the preceding chapter and it has been defined as the product of the geometrical distance between two points and the index of the medium between the points. Thus to a geometrical distance x measured in a medium of index μ the corresponding reduced distance is μx; but this assumes that the medium is uniform. For a heterogeneous medium the generalisation is clearly given by $\int \mu ds$, where ds is an element of the path between the limits of integration and μ is the index of the medium in which the element is measured. If the medium be uniform this clearly leads to the previous result. Thus the reduced distance, V, from a point P to a point P' is given by

$$V = \int_P^{P'} \mu ds.$$

Now from Chapter I, § 6, we have seen that the index of a medium is inversely proportional to the velocity of propagation of the luminous disturbance in that medium; whence it follows that the reduced distance, V, is proportional to the time taken by the disturbance to travel from the point P to the point P'. It is also proportional to the number of optical 'vibrations' between the two points, i.e., to the number of wave-lengths of the periodic disturbance.

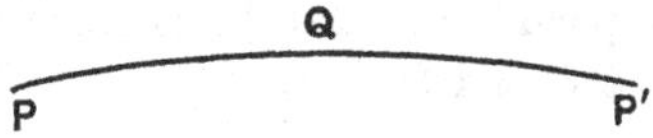

Let now a disturbance travel from a point P to a point P' by the path PQP', i.e., this path is at each point normal to the wave surface of the disturbance passing through that point. Then all the laws of geometrical optics are contained in the statement, due to Fermat, that the path PQP' is such that the time of propagation along it is stationary, i.e., for small variations of PQP'; or that the number of wave-lengths contained in it is stationary. These quantities may be either maxima or minima; the complete statement is

$$\delta \int_P^{P'} \mu ds = 0, \quad \text{.........................(1)}$$

the variation sign denoting a variation of the path PQP'. It may be verified quite readily that this leads to the usual laws of reflection and refraction of light; and by means of the Calculus of Variations we can obtain from it the differential equations of the path of the luminous disturbance in any medium in which μ is known at every point.

2. We assume that light is propagated under the form of a wave motion so that there will be wave surfaces or surfaces at every point of which the medium of propagation is in the same phase of disturbance; and the disturbance at every point will be travelling along the normal to the wave surface through that point. These normals in a homogeneous medium are, therefore, the paths of the luminous disturbance and take the place of the 'rays' previously considered. Now upon each normal there will be two points which are the centres of curvature of the two principal sections of the wave surface passing through that normal, and the locus of these points will be the surface of centres or caustic surface; so that each ray or normal touches the corresponding caustic in two points. We have seen that for any two points P and Q

$$\delta \int_P^{Q} \mu ds = 0, \quad \text{.............................(1)}$$

the path of integration being along the ray connecting these points; if however P and Q be upon corresponding caustic surfaces, then

$$\delta^2 \int_P^Q \mu ds = 0, \qquad \text{..........................(2)}$$

while in the particular case in which P and Q are cusps of these surfaces

$$\delta^3 \int_P^Q \mu ds = 0. \qquad \text{..........................(3)}$$

If we had perfect correspondence so that the wave surfaces were concentric spheres with centres at P and Q, then

$$\int_P^Q \mu ds = \text{const.},$$

i.e., the reduced path is the same for all rays connecting P and Q. The result (1) is equivalent to the statement that the variation of the reduced path from P to Q depends, to the first order of small quantities, only upon the displacements of P and Q themselves.

3. By an application* of the reduced path the relation between the two focal lengths of a symmetrical system may be proved—as foreshadowed in § 2 of the preceding chapter.

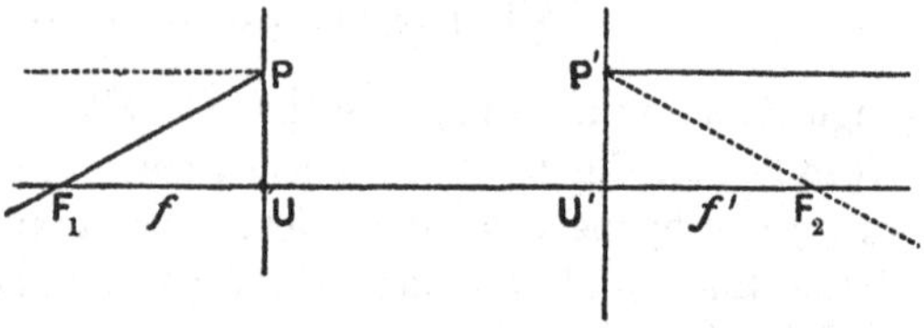

Let F_1 and F_2 be the principal foci of a symmetrical system and let any ray through F_1 intersect normal planes through the unit points in P and P' so that $UP = U'P' = h$, supposed small. Then in the usual notation $F_1U = f$ and $U'F_2 = f'$, giving the two focal lengths of the system. The emergent rays are parallel to the axis so that the optical length from F_1 to P' is independent of the position of P'; thus

$$V(F_1P') = V(F_1U')\dagger.$$

But $$F_1P = f(1 + h^2/2f^2) \text{ approximately,}$$

so that if μ and μ' be the optical indices of the two media

$$\mu f(1 + h^2/2f^2) + V(PP') = \mu f + V(UU'),$$

i.e., $$h^2\mu/2f = V(UU') - V(PP'). \qquad \text{..............(1)}$$

Similarly $$h^2\mu'/2f' = V(UU') - V(PP'),$$

* A. Pelletan, *Optique Appliquée*, chap. III.

† If aberrations be taken into consideration this equation is true up to, and including, terms in h^3.

so that equating these results

$$f/\mu = f'/\mu'. \quad \text{.........................(2)}$$

4. *The Characteristic Function.* We may apply the preceding considerations, of § 1, to any optical system. Let P and P' be two points upon two corresponding rays and let the co-ordinates of these points,

referred to two rectangular systems of axes, be x, y, z and x', y', z' respectively; let the direction cosines of the rays be L, M, N and L', M', N'. Then we may write

$$V = \int_P^{P'} \mu\, ds,$$

where the integration is taken along the ray and it is evident that V is a function of the variables x, y, z and x', y', z'. V is said to be the 'Characteristic Function *.' If we take two neighbouring points Q and Q', whose co-ordinates are $x + \delta x$, $y + \delta y$, $z + \delta z$ and $x' + \delta x'$, $y' + \delta y'$, $z' + \delta z'$, the value of the characteristic function for them may be written $V + \delta V$, and we have

$$V + \delta V = \int_Q^{Q'} \mu\, ds.$$

Thus

$$\delta V = \int_Q^{Q'} \mu\, ds - \int_P^{P'} \mu\, ds,$$

and owing to (1) § 1 the difference will arise from the change of P and Q to P' and Q'; projecting therefore upon the ray through P and its corresponding ray through P' we have, to the first approximation,

$$\delta V = -\mu L \delta x - \mu M \delta y - \mu N \delta z + \mu' L' \delta x' + \mu' M' \delta y' + \mu' N' \delta z'.$$

Whence
$$\left.\begin{aligned} -\mu L &= \frac{\partial V}{\partial x}, \quad -\mu M = \frac{\partial V}{\partial y}, \quad -\mu N = \frac{\partial V}{\partial z}, \\ \text{and} \quad \mu' L' &= \frac{\partial V}{\partial x'}, \quad \mu' M' = \frac{\partial V}{\partial y'}, \quad \mu' N' = \frac{\partial V}{\partial z'}. \end{aligned}\right\} \quad \text{.........(1)}$$

A knowledge of V, therefore, enables us to determine the initial and

* Introduced by Sir W. Hamilton; see references at the end of this chapter.

final directions of a ray passing through two given points. Moreover V satisfies the differential equations

$$\left(\frac{\partial V}{\partial x}\right)^2 + \left(\frac{\partial V}{\partial y}\right)^2 + \left(\frac{\partial V}{\partial z}\right)^2 = \mu^2,$$

$$\left(\frac{\partial V}{\partial x'}\right)^2 + \left(\frac{\partial V}{\partial y'}\right)^2 + \left(\frac{\partial V}{\partial z'}\right)^2 = \mu'^2.$$

If the optical system be symmetrical we shall find it convenient to restrict the movements of P and P' to planes normal to the axis of symmetry and then if this be taken as the x and x' axis in the two spaces we have only the final four relations of (1). It is evident that the transformation effected by the optical system is completely determined when the form of V is known, so that this function sums up in itself the properties of the system; and here V is a function of the four variables y, z, y' and z'.

5. *The Eikonal.* Let AA' be the axis of a symmetrical system and let P and P' be two points upon corresponding rays; draw normal planes through P and P' intersecting AA' in O and O' respectively—which may conveniently be taken as origins of co-ordinates: let AA' be the axis of x and of x' and take parallel rectangular axes in the two

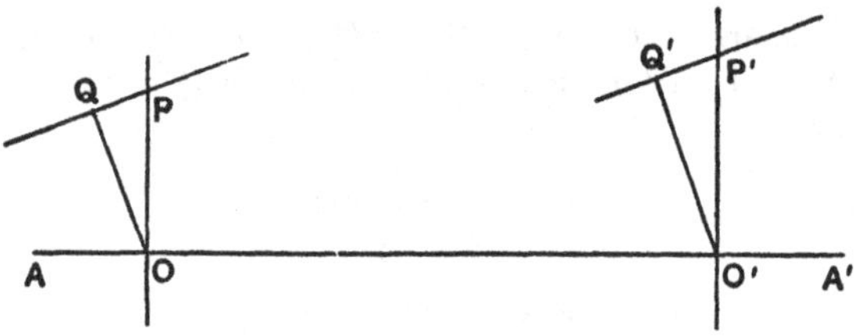

spaces. Let the co-ordinates of P and P' be y, z and y', z' respectively and write

$$V = \int_P^{P'} \mu\, ds$$

as usual; then V is a function of the four variables y, z, y' and z'. Let Q and Q' be the feet of perpendiculars from O and O' upon the ray. We may write

$$E = \int_Q^{Q'} \mu\, ds,$$

and then E is known as the 'Eikonal*.' Moreover we have

$$E = V + \mu(My + Nz) - \mu'(M'y' + N'z'), \quad \text{............(1)}$$

* The name 'Eikonal' was introduced by Bruns; see references. But the function appears much earlier in Hamilton's paper, *Theory of Systems of Rays*.

by projection of OP and $O'P'$ upon the two portions of the ray; where L, M, N and L', M', N' are the direction cosines of the ray in the two spaces. By variation of (1) we have

$$\delta E=\delta V+\mu(M\delta y+y\delta M+N\delta z+z\delta N)-\mu'(M'\delta y'+y'\delta M'+N'\delta z'+z'\delta N')$$
$$=\mu y\delta M+\mu z\delta N-\mu' y'\delta M'-\mu' z'\delta N'$$

from (1) § (4), since V is a function of y, z, y' and z';

$$\left.\begin{aligned}\text{i.e.,}\qquad \mu y&=\frac{\partial E}{\partial M}, & \mu z&=\frac{\partial E}{\partial N},\\ \text{and}\qquad -\mu' y'&=\frac{\partial E}{\partial M'}, & -\mu' z'&=\frac{\partial E}{\partial N'}.\end{aligned}\right\}\quad\ldots\ldots\ldots\ldots\ldots\ldots(2)$$

Thus we may regard E as a function of M, N, M' and N' alone and the properties of the system are completely known when the form of this function is known.

We may define also other functions as follows:

$$E_1=\int_Q^{P'}\mu\,ds \quad\text{and}\quad E_2=\int_P^{Q'}\mu\,ds,$$

and then by similar variations we obtain the relations

$$\mu y=\frac{\partial E_1}{\partial M},\quad \mu z=\frac{\partial E_1}{\partial N},\quad \mu' M'=\frac{\partial E_1}{\partial y'}\quad\text{and}\quad \mu' N'=\frac{\partial E_1}{\partial z'},$$

$$\text{and}\quad -\mu M=\frac{\partial E_2}{\partial y},\quad -\mu N=\frac{\partial E_2}{\partial z},\quad -\mu' y'=\frac{\partial E_2}{\partial M'}\quad\text{and}\quad -\mu' z'=\frac{\partial E_2}{\partial N'}.$$

Any of these functions may be used to investigate the properties of the optical system.

6. In the first place let us consider more fully the Eikonal function and write

$$e=\int_O^{O'}\mu\,ds,$$

where the integration is taken along the axis of the system. Further let U be defined by the relation

$$U=J(E-e),$$

where J is the modified power of the system; then using the relations of §11, Chapter I, we have

$$Y=\frac{\partial U}{\partial M},\quad Z=\frac{\partial U}{\partial N},\quad -Y'=\frac{\partial U}{\partial M'}\quad\text{and}\quad -Z'=\frac{\partial U}{\partial N'},\quad\ldots\ldots(1)$$

and U is known as the 'modified eikonal' or the 'eikonal expressed as a ratio.' It has the advantage that the 'size' of the system is eliminated, as also the explicit appearance of the optical indices of the various media. From (1) U is a function of the four variables M, N, M' and N', but owing to the symmetry of the system three variables only are sufficient,

viz., $$a = M^2 + N^2, \quad b = MM' + NN' \quad \text{and} \quad c = M'^2 + N'^2. \quad \text{...(2)}$$

We may therefore consider U as a function of these three variables a, b and c.

The base points chosen, O, O', may be any points whatever; in general we shall take them to be corresponding points, but one other case of considerable practical importance arises, viz., when they are the principal foci of the system. Then U is known as the 'focal reduced eikonal' and will be investigated later.

Let the base points chosen be corresponding points associated with a reduced magnification m; then if there were perfect definition over the two normal planes through O and O', associated with magnification m, we should have the relations

$$Y' - mY = 0 \quad \text{and} \quad Z' - mZ = 0,$$

i.e., $$\frac{\partial U}{\partial M'} + m\frac{\partial U}{\partial M} = 0 \quad \text{and} \quad \frac{\partial U}{\partial N'} + m\frac{\partial U}{\partial N} = 0,$$

for all values of the variables.

This suggests the use of the variables $M - mM'$ and $N - mN'$; in conformity with (2) above we choose three variables α, β and γ given in terms of a, b and c by the relations

$$\left.\begin{aligned} \alpha(s-m)^2 &= (M - sM')^2 + (N - sN')^2 = a - 2sb + s^2c, \\ \beta(s-m)^2 &= (M-sM')(M-mM') + (N-sN')(N-mN') = a-(s+m)b+smc, \\ \gamma(s-m)^2 &= (M - mM')^2 + (N - mN')^2 = a - 2mb + m^2c, \end{aligned}\right\} \quad \text{......(3)}$$

and owing to these linear relations U may be considered as a function of α, β and γ. Here s may be considered an arbitrary constant but it will be convenient to take it as the magnification associated with the axial points of the pupil planes; in any case it may be taken to be a magnification and we can draw the normal planes through the corresponding points which it defines upon the axis of the optical system. From Chapter I, §11, it is seen that $s - m$ is the distance in the image space between the points upon the axis defined by the magnifications s and m, i.e., the distance between the exit-pupil and the image plane. The factor $(s - m)^2$ is then added to the left-hand side of (3) for convenience in subsequent transformations.

7. From the relations (1) and (3) of §6 we have

$$(s-m)^2 Y' = -(s-m)^2 \frac{\partial U}{\partial M'}$$

$$= -(s-m)^2 \left\{\frac{\partial U}{\partial \alpha}\frac{\partial \alpha}{\partial M'} + \frac{\partial U}{\partial \beta}\frac{\partial \beta}{\partial M'} + \frac{\partial U}{\partial \gamma}\frac{\partial \gamma}{\partial M'}\right\}$$

$$= (M - sM')\left(2s\frac{\partial U}{\partial \alpha} + m\frac{\partial U}{\partial \beta}\right) + (M - mM')\left(s\frac{\partial U}{\partial \beta} + 2m\frac{\partial U}{\partial \gamma}\right),$$

and similarly

$$(s-m)^2 Y = (s-m)^2 \frac{\partial U}{\partial M}$$

$$= (M - sM')\left(2\frac{\partial U}{\partial \alpha} + \frac{\partial U}{\partial \beta}\right) + (M - mM')\left(\frac{\partial U}{\partial \beta} + 2\frac{\partial U}{\partial \gamma}\right).$$

By subtraction

$$(s-m)(Y' - mY) = 2(M - sM')\frac{\partial U}{\partial \alpha} + (M - mM')\frac{\partial U}{\partial \beta}.$$

Similarly

$$(s-m)(Z' - mZ) = 2(N - sN')\frac{\partial U}{\partial \alpha} + (N - mN')\frac{\partial U}{\partial \beta}. \quad \text{...(1)}$$

Now for perfect definition

$$Y' - mY = 0 \quad \text{and} \quad Z' - mZ = 0$$

for all values of Y and Z; thus either

$$\frac{\partial U}{\partial \alpha} = 0 \quad \text{and} \quad \frac{\partial U}{\partial \beta} = 0 \quad \text{......................(2)}$$

for every ray; or

$$\frac{M - sM'}{N - sN'} = \frac{M - mM'}{N - mN'} \quad \text{always,}$$

which implies that

$$\frac{M}{M'} = \frac{N}{N'}. \quad \text{.............................(3)}$$

This latter condition implies that all rays lie in a plane through the axis of the optical system; which clearly is not true. We are led back to the relation (2) and it follows that for perfect definition over the conjugate planes associated with magnification m, U must depend only upon the function γ, i.e., we may write

$$U = f(\gamma). \quad \text{.............................(4)}$$

8. *The Aberration Function.* In general there will not be perfect definition over the planes associated with magnification m; in Chapter I we saw that we were obliged to approximate to obtain the point-to-point geometrical correspondence with which we started. It follows that U will contain terms involving the variables α and β of § 6. Let us write

$$U = u - \Phi, \quad \text{.........................(1)}$$

where u contains only the variable γ, so that $u=f(\gamma)$. Φ consists of any other terms which may appear in the reduced eikonal, so that the existence of Φ implies the presence of aberrations and $\Phi=0$ only in a 'perfect' system, i.e., a system giving perfect definition over the conjugate planes at paraxial magnification m. For this reason Φ is called the 'Aberration Function.' It may be added that the negative sign is attached to Φ for subsequent formal simplicity; moreover $-\Phi$ corresponds with a function $+\Psi$ associated with the Characteristic Function.

$$\text{Now}\quad Y'-mY=-\left(\frac{\partial}{\partial M'}+m\frac{\partial}{\partial M}\right)(u-\Phi)=\left(\frac{\partial}{\partial M'}+m\frac{\partial}{\partial M}\right)\Phi,$$

since the operator considered annihilates any function of γ only; thus

$$Y'-mY=\frac{\partial\Phi}{\partial M'}+m\frac{\partial\Phi}{\partial M}\quad\text{and}\quad Z'-mZ=\frac{\partial\Phi}{\partial N'}+m\frac{\partial\Phi}{\partial N},\ldots(2)$$

and this exemplifies the fact that Φ gives completely the 'aberration' or wandering of the ray from ideal imagery.

We have written $u=f(\gamma)$ and so far the form of the function is arbitrary; it may therefore be determined to satisfy certain other conditions and this accordingly is considered below. Φ is a function of α, β and γ in general and we may write

$$\Phi=\sum_{n=2}\Phi_n,$$

where Φ_n is homogeneous and of degree n in the variables α, β and γ, since the aberrations are essentially of higher order than the first; to the first approximation we shall have ideal imagery. We are introduced here to the idea of 'orders of aberration'—which will be investigated more fully in the sequel.

9. *Axial Aberration for the Pupil Planes.* We have been considering the pair of points upon the axis defined by the magnification s and it is only as a first approximation that we may assume that corresponding to rays through one of these points there will be rays passing through the corresponding point. But we proceed to shew that the form of the function $f(\gamma)$ may be determined so that this is true accurately.

Let U_s be the reduced eikonal, the axial points of the pupil planes being base points associated with reduced magnification s; then

$$U_s=u+\frac{s-m}{sm}(L-1)-(s-m)(L'-1),\ \ldots\ldots\ldots\ldots(1)$$

assuming the system corrected for magnification m.

The points y, z and y', z' of intersection of the ray with the pupil planes are given by

$$\left.\begin{aligned} y &= \frac{\partial U_s}{\partial M} = \frac{2(M-mM')}{(s-m)^2}\frac{du}{d\gamma} - \frac{(s-m)}{sm}\frac{M}{L}, \\ y' &= -\frac{\partial U_s}{\partial M'} = \frac{2m(M-mM')}{(s-m)^2}\frac{du}{d\gamma} - (s-m)\frac{M'}{L'}, \end{aligned}\right\} \quad \text{.........(2)}$$

together with two similar equations for z and z'. If these co-ordinates are to vanish simultaneously, and the pupil planes be free from axial aberration, we must have

$$\frac{M}{L} = \frac{2sm(M-mM')}{(s-m)^3}\frac{du}{d\gamma} = \frac{sM'}{L'} \quad \text{..................(3)}$$

and similarly
$$\frac{N}{L} = \frac{2sm(N-mN')}{(s-m)^3}\frac{du}{d\gamma} = \frac{sN'}{L'}; \quad \text{..............(4)}$$

so that
$$2m(sL-mL')\frac{du}{d\gamma} = (s-m)^3. \quad \text{....................(5)}$$

Squaring and adding equations (3) and (4), and using (3) § 6,

$$\frac{1-L^2}{L^2} = s^2\gamma\theta^2 = s^2\frac{1-L'^2}{L'^2},$$

where $\theta(s-m)^2 = 2m\dfrac{du}{d\gamma}$: and these results may be substituted in (5) giving a biquadratic equation in θ, viz.,

$$\theta\left(\frac{s}{\sqrt{1+s^2\gamma\theta^2}} - \frac{m}{\sqrt{1+\gamma\theta^2}}\right) = s-m. \quad \text{..............(6)}$$

If $s^2=1$ the solution takes a particularly simple form, and this is a case of frequent occurrence; for then $\theta = (1-\gamma)^{-\frac{1}{2}}$ so that

$$u = -\frac{(s-m)^2}{m}(1-\gamma)^{\frac{1}{2}}. \quad \text{....................(7)}$$

In the more general case, for unrestricted values of s, we have to deal with (6), which may be written

$$\theta\left(1 + \sum_{n=1}^{\infty} \psi_n e_n \gamma^n \theta^{2n}\right) = 1,$$

where $\psi_n = {}_{-\frac{1}{2}}C_n$, the usual Binomial coefficient, and $e_n = (s^{2n+1}-m)/(s-m)$. By reversion of this series in the usual way we obtain θ in terms of γ

$$\theta = 1 + \tfrac{1}{2}e_1\gamma - \tfrac{3}{8}(e_2 - 2e_1^2)\gamma^2 + \tfrac{1}{16}(5e_3 - 24e_2e_1 + 24e_1^3)\gamma^3 + \dots$$

and
$$\frac{mu}{(s-m)^2} = \tfrac{1}{2}\int\theta d\gamma = \tfrac{1}{2}\gamma + \tfrac{1}{8}e_1\gamma^2 - \tfrac{1}{16}(e_2 - 2e_1^2)\gamma^3 + \tfrac{1}{128}(5e_3 - 24e_2e_1 + 24e_1^3)\gamma^4 + \dots,$$

the constant of integration having been omitted since u must vanish for the ray coinciding with the axis of the system. This then is the form of the eikonal for a system which is to be regarded as fully corrected for the magnification m*.

10. *The Characteristic Function; the Aberration Function.* The performance of the symmetrical optical system may be investigated also by means of the Characteristic Function V† of § 4, and we are led again to an Aberration Function; this form of potential function introduces considerable simplifications into the discussion of diffraction phenomena to be undertaken later and the following investigation is therefore given here.

In § 4 we have defined the characteristic function V by the relation

$$V = \int_P^{P'} \mu \, ds;$$

let us assume now, however, that O and O' are not corresponding points, but that they have associated with them magnifications m and s; in the sequel we shall take O' to be the centre of the exit-pupil. Then it is clear that V is a function of the four variables Y, Z, Y' and Z'; for distinction we may write y', z' in place of Y', Z', for it is desirable to keep these latter to denote the intersection of the ray with the normal plane through the point upon the axis conjugate to O. Owing to the symmetry of the system V will be a function of the three variables

$$Y^2 + Z^2, \quad Yy' + Zz', \quad y'^2 + z'^2;$$

and if we write $Y_1 = mY$ and $Z_1 = mZ$, so that Y, Z and Y_1, Z_1 represent corresponding points in an aberrationless system, our three variables may then conveniently be θ, ϕ and ψ given by the relations

$$\theta d^2 = Y_1^2 + Z_1^2, \quad \phi d^2 = 2(Y_1 y' + Z_1 z') \quad \text{and} \quad \psi d^2 = y'^2 + z'^2,$$

where d is the axial distance between the exit-pupil and the image plane. Moreover it is clear that the aberration is completely given by the expressions

$$Y' - Y_1 \quad \text{and} \quad Z' - Z_1.$$

Further, at any time we may assume without loss of generality that $Z_1 = mZ = 0$, by a proper choice of the axis of Z.

From the general property of the Characteristic Function

$$M' = \frac{\partial V}{\partial y'} \quad \text{and} \quad N' = \frac{\partial V}{\partial z'} \ddagger;$$

* Cf. T. Smith, *Trans. Opt. Soc.* XXIII, 1921–2, No. 5.

† For further information see a paper by the author on 'The Aberrations of a Symmetrical Optical System,' *Trans. Camb. Phil. Soc.* XXIII, No. 9, 1926.

‡ Using reduced distances y' and z'.

so that $$Y' = y' + \frac{M'd}{L'} \text{ and } Z' = z' + \frac{N'd}{L'}, \quad \text{...............(1)}$$

where the ray intersects the normal plane through the point conjugate to O at the point $Q'(Y', Z')$; then for an aberrationless system the reduced path from P to Q' measured along the ray will be the same for all rays, i.e., will depend only upon the co-ordinates of P or Q'; and $Y' = Y_1$ and $Z' = Z_1$. Thus

$$\int_P^{Q'} \mu\, ds = f(\theta),$$

where $f(\theta)$ is an undetermined function of θ. Moveover

$$\int_{P'}^{Q'} \mu\, ds = \{(Y_1 - y')^2 + (Z_1 - z')^2 + d^2\}^{\frac{1}{2}} = d\sqrt{1 + \theta - \phi + \psi} = d\sqrt{1 + u},$$

where $u = \theta - \phi + \psi$. Therefore we have

$$V = \int_P^{P'} \mu\, ds = \int_P^{Q'} \mu\, ds - \int_{P'}^{Q'} \mu\, ds = f(\theta) - d\sqrt{1 + u}. \quad \text{......(2)}$$

Any terms appearing in V other than these shewn in (2) will involve aberrations and we may write

$$V = f(\theta) - d\sqrt{1 + u} + F, \quad \text{..................(3)}$$

where F is in general a function of the three variables θ, ϕ and ψ and denotes the presence of aberrations. F is called an Aberration Function and, as in the case of § 8,

$$F = 0$$

denotes a system giving perfect definition over the conjugate planes associated with magnification m; or alternatively F gives completely the aberrations of the system for these planes.

11. In confirmation of this we may combine formulae (1) and (2) above; thus, assuming $F = 0$,

$$M'd^2 = d^2 \frac{\partial V}{\partial y'} = 2Y_1 \frac{\partial V}{\partial \phi} + 2y' \frac{\partial V}{\partial \psi} \text{ and } N'd^2 = 2Z_1 \frac{\partial V}{\partial \phi} + 2z' \frac{\partial V}{\partial \psi}\text{*},$$

i.e., $M'd^2 = d(Y_1 - y')/\sqrt{1 + u}$ and $N'd^2 = d(Z_1 - z')/\sqrt{1 + u}$;

whence $$L'^2 = 1 - M'^2 - N'^2 = 1 - \frac{u}{1 + u} = \frac{1}{1 + u}$$

and $$Y' = y' + (Y_1 - y') \frac{\sqrt{1 + u}}{\sqrt{1 + u}},$$

i.e., $$Y' = Y_1 \text{ and similarly } Z' = Z_1.$$

* We assume henceforth that V has been multiplied by J, the modified power of the system, so that Y', Z', y' and z' are modified and reduced distances; and $d = s - m$.

It will be seen that the partial differentiations employed above do not involve the variable θ and the function $f(\theta)$ is arbitrary; it may therefore be determined so as to satisfy other conditions and the conditions chosen are the point-to-point correspondence of the axial points of the pupil planes.

12. *Elementary Properties.* In § 6 we have taken base points O and O' which are conjugate axial points associated with magnification m; if now we take new base points O_1 and O_1' associated with magnification m_1 and denote by the suffix unity the corresponding eikonal and aberration function we shall have new variables α_1, β_1 and γ_1, in accordance with (3) §6; and we may write down the relation between $(u-\Phi)$ and $(u-\Phi)_1$*. It is clear then that $\Phi=0$ does *not* imply $\Phi_1=0$: thus if we have perfect definition over the planes m we shall not necessarily have perfect definition over any other pair of conjugate normal planes. In other words a symmetrical optical system is designed in general for only one position of the object or of the image.

But if we consider only the first powers of the variables the aberration function will not appear; and then we have perfect definition—to this degree of approximation—over all normal planes; and it is of interest to verify the elementary properties of the system by the use of the eikonal function. Thus

$$U=u=f(\gamma)=\frac{1}{2m}\{(M-mM')^2+(N-mN')^2\};$$

and $$Y=\frac{\partial U}{\partial M}=\frac{1}{m}(M-mM'),\qquad Y'=-\frac{\partial U}{\partial M'}=M-mM';$$

so that $Y'=mY$ and $Z'=mZ$. Further, if $m=0$,

$$Y'=M \text{ and similarly } Z'=N;$$

all rays parallel before refraction pass therefore, on emergence, through a point upon the plane defined by $m=0$; in particular if $M=N=0$, i.e., the incident rays are parallel to the axis of the system, they will pass afterwards through an axial point, viz., the second principal focus. Similarly for the first principal focus and focal plane. Again, we may define the unit planes by the relation† $\mu m=\mu'$; in which case

$$Y'/Y=m=\mu'/\mu,\quad Z'/Z=m=\mu'/\mu,$$

so that the geometrical sizes of object and image are the same. If $m=1$, $Y=0$, and $Z=0$, we have $Y'=0$, $Z'=0$, and also

$$M=M',\quad N=N',$$

* See Chapter V.

† m is the reduced magnification of § 11, Chapter I.

i.e., we have defined the nodal points. Further, in general, for fixed axial points

$$M - mM' = 0 = N - mN',$$

i.e., in the usual notation

$$\mu l \sin a = \text{const.}, \quad \text{.........................(1)}$$

or as an approximation to this

$$\mu l a = \text{const.},$$

as given by Robert Smith* and Helmholtz. More generally

$$M - mM' = mY = Y_1, \quad N - mN' = mZ = Z_1, \quad \text{......(2)}$$

so that for all rays through a given object point these expressions are constant; and this 'cosine law' is an extension of the sine-relation (1) into the outer parts of the field. It will be considered more fully in a subsequent chapter.

The relations (2) may be obtained in a more general way; thus if we assume perfect definition over the conjugate planes m we shall have $\Phi = 0$ and then $U = f(\gamma)$, so that

$$Y = \frac{\partial U}{\partial M} = \frac{2(M - mM')}{d^2}\frac{\partial f}{\partial \gamma}, \quad Z = \frac{\partial U}{\partial N} = \frac{2(N - mN')}{d^2}\frac{\partial f}{\partial \gamma}:$$

i.e., $$Y/Z = (M - mM')/(N - mN') \text{ and } Y^2 + Z^2 = 4\gamma\left(\frac{\partial f}{\partial \gamma}\right)^2.$$

From these it follows that $M - mM'$, $N - mN'$ and γ depend only upon the particular point of the object from which the ray originates.

REFERENCES

Bruns, H. Das Eikonal. *Saechs. Ber. d. Wiss.* XXI (1895).

Hamilton, Sir W. R. *The Mathematical Papers of Sir William Rowan Hamilton*, vol. I. *Geometrical Optics* (Cambridge, 1931).

Klein, F. *Ueber das Brunsche Eikonal.*

Maxwell. On the Application of Hamilton's Characteristic Function to Optical Instruments.... *Proc. L. M. S.* VI (1874–5).

Moebius, A. F. *Leipziger Berichte*, VII (1855).

Smith, T. *Trans. Opt. Soc.* XXIII (1921–2), No. 5.

—— *Dictionary of Applied Physics*, Art. 'Optical Calculations.'

—— *Proc. Opt. Conv.* 1926, Part II, p. 740.

Steward, G. C. *Phil. Trans.* A, 225 (1925), Part I.

—— *Trans. Camb. Phil. Soc.* XXIII, No. 9 (1926).

—— *Proc. Opt. Conv.* 1926, Part II, p. 778.

Thiessen. Beiträge zur Dioptrik. *Berl. Berichte*, 1890.

* *Compleat Opticks*, Cambridge, 1738.

CHAPTER III

THE GEOMETRICAL ABERRATIONS

1. *General Properties.* We have seen in the previous chapter that the aberrations of a symmetrical optical system are given completely by an aberration function and we have already defined two forms of this function, viz., the Φ of § 8 and the F of § 10: let us consider for a moment the former of these functions. The wandering of the ray from the ideal focus is given by the relations

$$Y' - mY = \frac{\partial\Phi}{\partial M'} + m\frac{\partial\Phi}{\partial M} \text{ and } Z' - mZ = \frac{\partial\Phi}{\partial N'} + m\frac{\partial\Phi}{\partial N}. \quad \text{...(1)}$$

Now Φ is a function of the three variables α, β and γ introduced in § 6, Chapter II, and inasmuch as the aberrations belong to the second and higher approximations we may write

$$\Phi = \sum_{n=2}^{\infty} \Phi_n, \quad \text{.............................(2)}$$

where Φ_n is a homogeneous function of degree n in the variables α, β and γ. Each term in Φ will give rise to 'an aberration' and the group of terms denoted by Φ_n will give aberrations of one particular order, viz., of order $n-1$; for it is convenient to name those corresponding to the lowest term in (2), Φ_2, the 'first order' or 'primary' aberrations. The number of terms in Φ_n will be $\frac{1}{2}(n+1)(n+2)$, so that there is apparently this number of aberrations of order $(n-1)$; but any term containing γ only is annihilated by the operator (1) above, and one such term will appear in every function Φ_n. Thus the total number of aberrations of order $(n-1)$ is $\frac{1}{2}n(n+3)$*; we have therefore *five* of the first order, *nine* of the second order, *fourteen* of the third order and so on. These five primary aberrations are the so-called 'five aberrations of von Seidel' †; but in this connection it is of interest to note that all five of them were discussed by Herschel, Airy, Coddington and Hamilton ‡ before the time of von Siedel, and to Airy is due also the Petzval' condition relating to the curvature of images—as far at all events as it refers to the combination of thin lenses. This condition will be examined in detail subsequently.

2. *Change of Focus.* The formulae (1) § 1 give the departures from ideal imagery upon the Gaussian conjugate plane; it is important to consider also the intersection of the emergent ray with neighbouring

* Rayleigh, *Collected Works*, v, p. 453. † *Ibid.* p. 456.

‡ Cf. especially 'On the Optical Writings of Sir William Rowan Hamilton', *Math. Gazette*, July 1932, vol. XVI, no. 219.

planes so that out-of-focus effects may be investigated. Let E be the axial point of the exit-pupil and P_1 the ideal image upon the Gaussian plane $O'P_1P'$ and let an emergent ray $Q'P'$ cut the pupil plane in Q' and a normal plane distant X from O' in P''; X is supposed small. Let EP_1 intersect this plane in E_1; EP_1 will be named the 'central line': let P_1' be the orthogonal projection of P_1 upon the normal plane $O''E_1P''$.

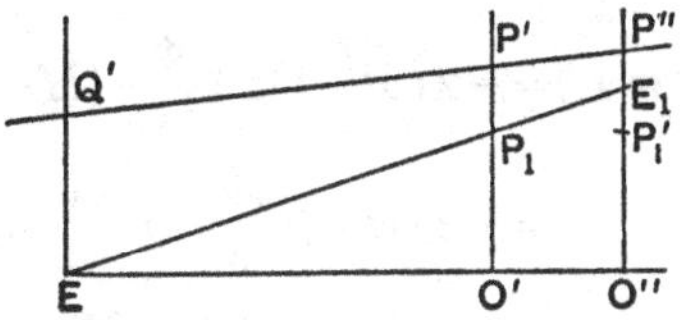

In conformity with the previous notation the co-ordinates of P_1 will be Y_1, Z_1; of P', Y', Z'; of Q', y', z' and those of P'', Y'', Z'': and $O'O''=X$, $EO'=d=s-m$. Then

$$d(Y''-Y')=X(Y'-y'),\ d(Z''-Z')=X(Z'-z');$$

so that

$$d(Y''-Y_1)=d(Y'-Y_1)+X(Y_1-y'),$$

$$d(Z''-Z_1)=d(Z'-Z_1)+X(Z_1-z'), \quad \text{............(1)}$$

since X, $Y'-Y_1$ and $Z'-Z_1$ are small. These equations give the aberration displacements upon the out-of-focus plane X and we may omit the terms XY_1 and XZ_1 provided that we change our origin from P_1' to E_1 upon the central line.

3. *The Primary Displacement.* Considering now the function Φ let us examine in detail the first order aberrations: we write

$$\Phi_2=\tfrac{1}{8}\{\sigma_1\alpha^2-4\sigma_2\alpha\beta+2\sigma_3\alpha\gamma+4\sigma_4\beta^2-4\sigma_5\beta\gamma+\sigma_6\gamma^2\}, \quad \text{...(1)}$$

the variables being those of § 6, Chapter II. The numerical coefficients appearing here are inserted for subsequent simplicity; they are suggested by the expansion of the expression $(a-2b+c)^2$ which occurs in the focal eikonal to be evaluated subsequently. The coefficients $\sigma_1, \ldots \sigma_6$ are known as 'aberration coefficients' and as will appear shortly they control completely the first order aberrations of the system.

Now to the first approximation we may neglect aberrations and we have

$$u=\frac{(s-m)^2}{2m}\gamma;$$

so that

$$Y=\frac{\partial u}{\partial M}=\frac{1}{m}(M-mM') \text{ and similarly } Z=\frac{1}{m}(N-mN').$$

Similarly $\quad \rho\cos\phi=M-sM'$ and $\rho\sin\phi=N-sN'$, $\quad$ ....(2)

ρ and ϕ being polar co-ordinates of a point upon the exit-pupil given by the magnification s; so that

$$\alpha d^2 = \rho^2, \quad \beta d^2 = \rho Y_1 \cos\phi, \quad \gamma d^2 = Y_1^2, \quad \text{.............. (3)}$$

where $Z_1 = 0$.

From §1 therefore we have

$$Y' - mY = \left(\frac{\partial}{\partial M'} + m\frac{\partial}{\partial M}\right)\Phi = \left(\frac{\partial \alpha}{\partial M'} + m\frac{\partial \alpha}{\partial M}\right)\frac{\partial \Phi}{\partial \alpha} + \left(\frac{\partial \beta}{\partial M'} + m\frac{\partial \beta}{\partial M}\right)\frac{\partial \Phi}{\partial \beta},$$

i.e.,

$$d(Y' - mY) = -2(M - sM')\frac{\partial \Phi}{\partial \alpha} - (M - mM')\frac{\partial \Phi}{\partial \beta}$$

$$= -2\rho\cos\phi\frac{\partial \Phi}{\partial \alpha} - Y_1\frac{\partial \Phi}{\partial \beta}, \quad \text{.................. (4)}$$

and similarly

$$d(Z' - mZ) = -2\rho\sin\phi\frac{\partial \Phi}{\partial \alpha}.$$

Substituting, therefore, and using (1), (2) and (3) we have

$$\left.\begin{aligned} 2d^3(Y' - mY) &= \sigma_1 S + \sigma_2 C + (\sigma_3 + 2\sigma_4)A + \sigma_5 D, \\ 2d^3(Z' - mZ) &= \sigma_1 S' + \sigma_2 C' + \sigma_3 A', \end{aligned}\right\} \quad \text{...... (5)}$$

where

$$\begin{matrix} S \\ S' \end{matrix} \equiv -\rho^3 \begin{matrix} \cos\phi \\ \sin\phi \end{matrix}; \quad \begin{matrix} C \\ C' \end{matrix} \equiv \rho^2 Y_1 \begin{matrix} (2 + \cos 2\phi) \\ \sin 2\phi \end{matrix}; \quad \begin{matrix} A \\ A' \end{matrix} \equiv -\rho Y_1^2 \begin{matrix} \cos\phi \\ \sin\phi \end{matrix}; \quad D \equiv Y_1^3.$$

Equations (5) give completely the departure from ideal imagery as far as the first order aberrations are concerned.

4. *Spherical Aberration.* Let us consider the first coefficient σ_1 appearing in (5) § 3; we have

$$Y' - mY = -\frac{\sigma_1}{2}\left(\frac{\rho}{d}\right)^3 \cos\phi, \quad Z' - mZ = -\frac{\sigma_1}{2}\left(\frac{\rho}{d}\right)^3 \sin\phi. \quad \text{... (1)}$$

It is evident that if we write $Y_1 = 0$, i.e., if we consider an object point upon the axis of symmetry, this term in σ_1 is the only one remaining in (5) §3; so that this is the only first order axial aberration. We may write here, and generally, also $Z_1 = 0$; then (1) indicates that the ray cuts the image plane through O' upon the circumference of a circle, centre O' and radius $\frac{\sigma_1}{2}\left(\frac{\rho}{d}\right)^3$, so that for a vanishing ρ the rays will pass through O': for any other value of ρ they will intersect the axis at some point O_1 where from the figure $O_1O' = \frac{\sigma_1}{2}\left(\frac{\rho}{d}\right)^2$. This is called the 'longitudinal' aberration; $O'P'$ is the 'lateral' aberration. Moreover, in general, the rays will all touch a 'caustic' surface of revolution obtained from the curve

$$8x^3 + 27\sigma_1 y^2 = 0, \quad \ldots\ldots\ldots\ldots(2)$$

where O' is the origin of co-ordinates and the axis of x the axis of the system. Let an extreme ray cut the caustic surface in Q_0 and draw NQ_0

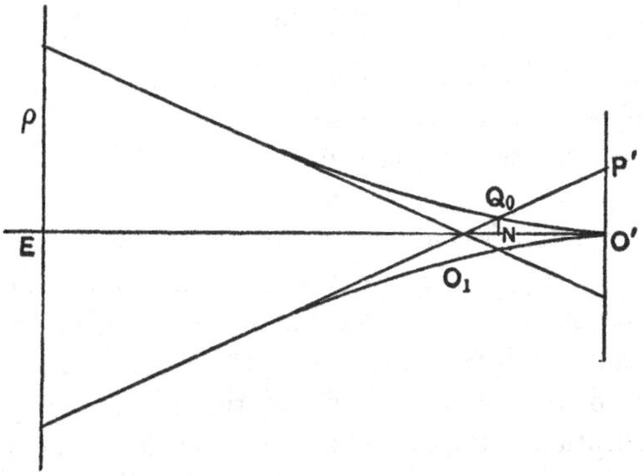

normal to the axis; then all rays intersect the normal plane through N within a circle of radius NQ_0 of magnitude $\frac{\sigma_1}{8}\left(\frac{\rho}{d}\right)^3$, so that $3O_1N = NO'$. It is clear that this circle is the smallest circular area of illumination for any position of the receiving plane and there has been attached to it, therefore, by writers on geometrical optics, a special significance from the point of view of focussing.

Thus all the properties of the figure depend only upon the coefficient σ_1 and this aberration is known as 'spherical aberration.' We have assumed an incident beam free from aberration, but this is not necessary; for otherwise we have merely the increment in aberration and it may be verified immediately, using the results of Chapter v, that the above result giving the longitudinal aberration leads at once to the formulae usually given*.

Again we have written $Y_1 = 0$ and this is not necessary; for if $Y_1 \neq 0$ the whole investigation applies as above, in the absence of the other aberrations, except that the central line is now taken instead of the axis of the system.

5. *Coma.* We consider next the terms in σ_2 for which we have

$$Y' - mY = \frac{\sigma_2}{2}\frac{\rho^2 Y_1}{d^3}(2 + \cos 2\phi), \quad Z' - mZ = \frac{\sigma_2}{2}\frac{\rho^2 Y_1}{d^3}\sin 2\phi; \ldots(1)$$

* Herman. *Geometrical Optics* chap. VIII.

so that, if we take now our origin at the point P_1, this indicates that corresponding to an annulus of the exit-pupil of radius ρ we have upon the image plane a circle of radius $\frac{\sigma_2}{2}\left(\frac{\rho}{d}\right)^2\left(\frac{Y_1}{d}\right)$, the centre being at the point $\sigma_2\left(\frac{\rho}{d}\right)^2\left(\frac{Y_1}{d}\right)$, 0. For variations of ρ both centre and radius change; but it is clear that all the circles touch two straight lines through P_1, symmetrically placed, and inclined to one another at an angle of 60°. All the illumination therefore is contained within this angle, being more intense towards the head P_1; so that we have the comet appearance or 'coma' flare. This aberration accordingly is known as 'coma,' or better 'circular coma,' to distinguish it from what follows*. From the occurrence of 2ϕ in (1) it is evident that a single description of the exit-pupil implies a double description of the corresponding coma circle, and also that rays from diametrically opposite points of the exit-pupil intersect upon this circle, i.e., upon the Gaussian conjugate plane. This implies that change of focus will be of no benefit with this aberration. We have also

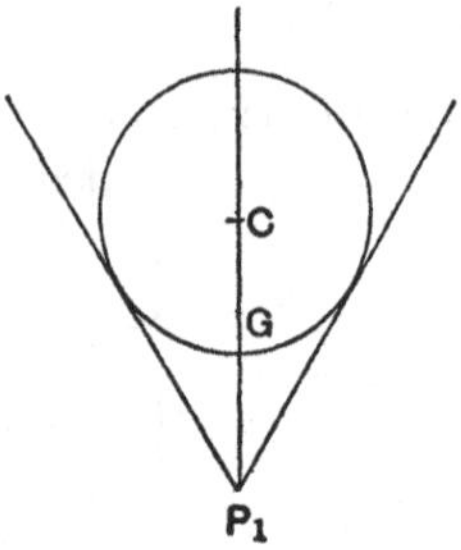

$$P_1 G = \frac{\sigma_2}{2}\left(\frac{\rho}{d}\right)^2\left(\frac{Y_1}{d}\right), \quad \text{.......................(2)}$$

where C is the centre of a coma circle which is met by P_1C in G. The coefficient σ_2 therefore governs completely the first order circular coma.

6. *Astigmatism and Curvature of the Field.* The two coefficients σ_3 and σ_4 occur together in (5) § 3; we have

$$Y' - mY = -\tfrac{1}{2}(\sigma_3 + 2\sigma_4)\frac{\rho}{d}\left(\frac{Y_1}{d}\right)^2\cos\phi, \quad Z' - mZ = -\tfrac{1}{2}\sigma_3\frac{\rho}{d}\left(\frac{Y_1}{d}\right)^2\sin\phi; \quad \text{......(1)}$$

these equations indicate an elliptical displacement from the Gaussian conjugate, the axes of the ellipse being proportional to $(\sigma_3 + 2\sigma_4)$ and σ_3 respectively. Let us consider, in the first place, rays in an axial plane passing through the object point Y, i.e., $\phi_1 = 0$ or π.

If P_1 be the Gaussian conjugate and E_1 the centre of the exit-pupil the first of the relations (1) indicates that the two rays corresponding to $\phi = 0$ and π intersect in a point F_1 upon the central line E_1P_1; the

* Viz., 'elliptical coma' in §11 and 'generalised coma' in §13.

aggregate of such points F_1 for various positions of the object point will give a surface. Moreover if x be the distance of F_1 from the conjugate plane we have

$$\frac{x}{d-x} = -\tfrac{1}{2}(\sigma_3 + 2\sigma_4)\frac{1}{d}\left(\frac{Y_1}{d}\right)^2,$$

i.e., $$x = -\tfrac{1}{2}(\sigma_3 + 2\sigma_4)\left(\frac{Y_1}{d}\right)^2 \text{ (approximately)}, \quad \text{........(2)}$$

since x is small; the negative sign indicates that F_1 is in front of the conjugate plane for positive values of σ_3 and σ_4. Relation (2) shews that

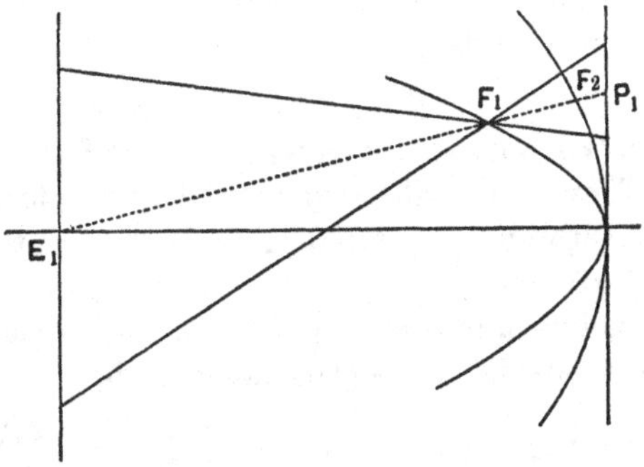

the locus of F_1 is a surface of revolution about the axis of the system and that the curvature of this surface is $(\sigma_3 + 2\sigma_4)/d^2$; but this is the modified and reduced curvature. We have assumed here that the object surface is plane; if however it be a surface of revolution then the expression just obtained is the increment in curvature of the image surface; so that if R_1 be this latter curvature we have

$$\Delta(R_1) = (\sigma_3 + 2\sigma_4)/d^2. \quad \text{....................(3)}$$

The surface which we have just obtained is known as the 'primary' surface; we may similarly obtain a 'secondary' surface by considering rays for which $\phi = \frac{\pi}{2}$ or $\frac{3\pi}{2}$; these will intersect in a point F_2 upon E_1P_1 and the aggregate of such points will give the secondary surface. If R_2 be the curvature of this, we have

$$\Delta(R_2) = \sigma_3/d^2. \quad \text{..........................(4)}$$

Moreover from (2) the separation of the surfaces is given by the expression

$$[x] = \sigma_4\left(\frac{Y_1}{d}\right)^2. \quad \text{..........................(5)}$$

Now we may write (3) and (4) as follows:

$$\Delta(R_1) = 3\sigma_4/d^2 + \varpi, \quad \text{......................(6)}$$

and

$$\Delta(R_2) = \sigma_4/d^2 + \varpi, \quad \text{......................(7)}$$

where $\varpi d^2 = \sigma_3 - \sigma_4$.

In the general case this aberration is known as 'astigmatism'; from (5) if $\sigma_4 = 0$ the surfaces coincide and we have simple 'curvature of the field': i.e., a normal plane in the object space is transformed into a surface of revolution. From this it will be seen that, in general, the so-called 'cardinal planes' are not plane at all but, instead, surfaces of revolution. If in addition $\varpi = 0$ the surface becomes a plane; this, therefore, is the condition for flatness of field, i.e.,

$$\varpi d^2 = \sigma_3 - \sigma_4 = 0; \quad \text{.........................(8)}$$

this is known as 'Petzval's condition for flatness of field,' and ϖ is the 'Petzval Sum.' These results will be evaluated more fully for actual optical systems subsequently; we are concerned here only with the qualitative nature of the aberrations.

From §2 the points of intersection of the ray with a plane a distance X in front of the paraxial image plane are given by

$$\eta = \left\{X - \tfrac{1}{2}(\sigma_3 + 2\sigma_4)\left(\frac{Y_1}{d}\right)^2\right\}\frac{\rho}{d}\cos\phi, \quad \zeta = \left\{X - \tfrac{1}{2}\sigma_3\left(\frac{Y_1}{d}\right)^2\right\}\frac{\rho}{d}\sin\phi, \quad \text{......(9)}$$

where η, ζ are co-ordinates referred to the 'central point' E_1 of §2 as origin. Now (9) will represent a circle, the 'least circle of confusion,' if

$$2Xd^2 = (\sigma_3 + \sigma_4)\,Y_1^2; \quad \text{....................(10)}$$

otherwise the area of illumination will be elliptical. Here the receiving plane bisects F_1F_2 and this accordingly will be the position of best focus according to the geometrical theory. Again, if either of the bracket terms in (9) vanish the area of illumination will reduce to a straight line—the 'focal line'; and this occurs when the receiving plane passes through F_1 or F_2. It is clear from (10) that the centres of the various circles of least confusion, corresponding to different positions of the object, will all lie upon a surface of revolution of curvature $(\sigma_3 + \sigma_4)/d^2$ and touching the two 'image' surfaces in their common axial points.

Again, it appears that astigmatism and curvature of the field are governed completely by the two aberration coefficients σ_3 and σ_4.

7. *Distortion.* We have finally to consider the coefficient σ_5; from this we obtain

$$Y' - mY = \tfrac{1}{2}\sigma_5(Y_1/d)^3, \quad Z' - mZ = 0. \quad \text{...........(1)}$$

These relations indicate a motion of the representative point from the ideal image away from, or towards, the axis of the system according as σ_5 be positive or negative; and further the magnitude of the motion varies as the cube of Y_1. Let us consider the image of a network α placed normal to the axis of the system; if σ_5 be positive the image is as

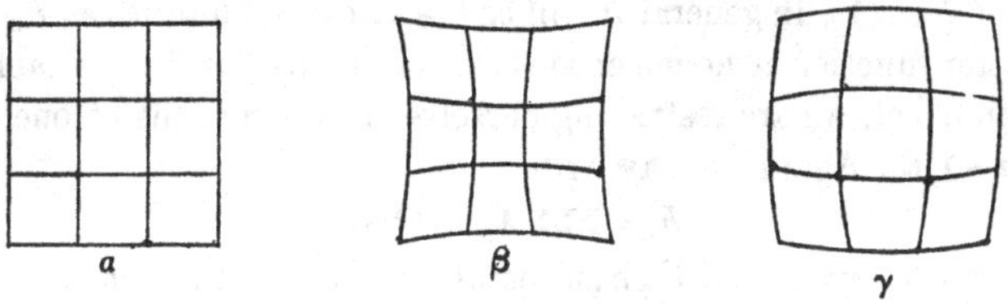

β while if σ_5 be negative we have the case of γ. We are dealing here with 'distortion' and it is known as 'pin-cushion' or 'barrel' distortion according as it corresponds to β or to γ, i.e., according as σ_5 is positive or negative.

It has appeared from the preceding discussion that there are *five* aberrations of the first order—the so-called 'five aberrations of von Seidel'; and that they are given completely by the five 'aberration coefficients' σ_1, σ_2, σ_3, σ_4 and σ_5. Lord Rayleigh has remarked that they do not stand all upon the same level; for three of them refer to errors of focussing (spherical aberration, coma and astigmatism) while two refer to the position of the focus (curvature of the field and distortion).

8. *Aberrations of Higher Orders.* We may deal with the aberrations of higher orders after the manner of the preceding paragraphs, for we may use in succession the functions Φ_3, Φ_4, ... Φ_n, ...; but we propose now to use the Characteristic Function, partly as an example of its use and partly because of its greater convenience in dealing with diffraction phenomena. In the notation already introduced V, and also the aberration function F, are functions of the variables θ, ϕ and ψ; and we have the relation

$$V = f(\theta) - d\sqrt{1+u} + F. \quad \text{.....................(1)}$$

Moreover the inclination of the emergent ray to the axis is given by

$$M' = \frac{\partial V}{\partial y'} \text{ and } N' = \frac{\partial V}{\partial z'};$$

so that

$$M'd^2 = 2\left(Y_1\frac{\partial V}{\partial \phi} + y'\frac{\partial V}{\partial \psi}\right) \text{ and } N'd^2 = 2\left(Z_1\frac{\partial V}{\partial \phi} + z'\frac{\partial V}{\partial \psi}\right). \quad \text{...(2)}$$

Again, $\quad Y' = y' + M'd/L'$ and $Z' = z' + N'd/L'$;

so that, substituting from (1) in (2), we have

$$d(Y'-Y_1)=2\left(Y_1\frac{\partial F_n}{\partial\phi}+y'\frac{\partial F_n}{\partial\psi}\right)\text{ and }d(Z'-Z_1)=2\left(Z_1\frac{\partial F_n}{\partial\phi}+z'\frac{\partial F_n}{\partial\psi}\right), \quad\ldots\ldots(3)$$

retaining only the terms of lowest order; where we have written F_n in place of F in (1). In general F will be the sum of such terms as F_n where the latter function is homogeneous and of degree n in the variables, so that, in effect, we are restricting ourselves to aberrations of one order, the $(n-1)$th. Again we may write

$$F_n=\Sigma\Sigma\Sigma A_{p,q,r}\,\theta^p\phi^q\psi^r,$$

where $p+q+r=n$, and then substituting in (3) and remembering that the polar co-ordinates ρ, ϕ_1* correspond to y', z' we have

$$\left.\begin{aligned} Y'-Y_1&=2^qA_{p,q,r}\left(\frac{Y_1}{d}\right)^{2p+q}\left(\frac{\rho}{d}\right)^{q+2r-1}\cos^{q-1}\phi_1\{2r\cos^2\phi_1+q\},\\ Z'-Z_1&=2^qA_{p,q,r}\left(\frac{Y_1}{d}\right)^{2p+q}\left(\frac{\rho}{d}\right)^{q+2r-1}2r\cos^q\phi_1\sin\phi_1,\end{aligned}\right\}\ldots(4)$$

corresponding to the term $A_{p,q,r}$, where we have written $Z_1=0$, as is always legitimate. These formulae (4) give completely the aberration displacements of order $(n-1)$ exhibited in powers of Y_1 and ρ, and taking $n=2$ it may be verified directly that they agree with the results of the preceding sections. In fact if we write, for the first order aberrations,

$$F_2=a_1\theta^2+a_2\theta\phi+a_3\theta\psi+a_4\phi^2+a_5\phi\psi+a_6\psi^2,$$

the coefficients here correspond to the σ-coefficients of the eikonal; and we have $\sigma_1+8a_6=0$, $\sigma_2=4a_5$, $\sigma_3+4a_3=0$, $\sigma_4+8a_4=0$ and $\sigma_5=4a_2$.

9. It is not proposed to examine here the higher order aberrations in detail; for this reference may be made elsewhere†; but only such of them as present peculiarities. And for those of the second order we may write

$$F_3=b_1\theta^3+b_2\theta^2\phi+b_3\theta^2\psi+b_4\theta\phi^2+b_5\theta\phi\psi+b_6\theta\psi^2+b_7\phi^3+b_8\phi^2\psi+b_9\phi\psi^2+b_{10}\psi^3.$$

The coefficient b_{10} gives spherical aberration, depending however upon the fifth power of the radius of the exit-pupil; but this type of aberration will follow whenever ψ appears alone. Consider a term $A_n\psi^n$ appearing in F; then from (4) §8 we have

$$d^{2n-1}(Y'-Y_1)=2nA_n\rho^{2n-1}\cos\phi_1,\quad d^{2n-1}(Z'-Z_1)=2nA_n\rho^{2n-1}\sin\phi_1*;$$

* ϕ_1 replaces the ϕ of §§ 1–7: it is used for distinction from the function ϕ of Chapter II, § 10, introduced with the Characteristic Function.

† *Trans. Camb. Phil. Soc.* XXIII, No. 9, 1926.

and this clearly repeats the investigation of §4. It is seen, therefore, that for spherical aberration of order n, the lateral aberration depends upon the $(2n+1)$th power of ρ while the longitudinal aberration depends upon the $2n$th power of ρ; and for an axial object point this is the only type of aberration which will appear.

Taking now the term in b_9, this represents circular coma, but it may be investigated for the general case; for if we have a term $A_{n+1}\phi\psi^n$ appearing in the aberration function F this will give, on substitution in (4) § 8,

$$d^{2n+1}(Y'-Y_1) = 2A_{n+1}Y_1\rho^{2n}(n\cos 2\phi_1 + n + 1),$$
$$d^{2n+1}(Z'-Z_1) = 2n\,A_{n+1}Y_1\rho^{2n}\sin 2\phi_1,$$

and this repeats the investigation of § 5. The displacement curve is as in the diagram of that paragraph but the angle between the tangents is no longer 60° but $2\sin^{-1} n/(n+1)$, the length P_1G being given by the relation

$$d^{2n+1}P_1G = 2A_{n+1}Y_1\rho^{2n}.$$

This aberration may be named 'circular' coma of order n, to distinguish it from other types of coma, to be examined later, in which the displacement curve is no longer a circle*.

10. The coefficients b_6 and b_8 fall naturally together and from (4) § 8 we have

$$\left.\begin{aligned}\eta &= (\cos^2\phi_1 + 1 + a)\cos\phi_1,\\ \zeta &= (\cos^2\phi_1 + a)\sin\phi_1,\end{aligned}\right\} \qquad (1)$$

where $8b_8\rho^3Y_1^2\eta = d^5(Y'-Y_1)$ and a similar expression for ζ and Z; and $2ab_8 = b_6$. Thus η and ζ are current co-ordinates referred to the ideal image point as origin and (1) gives the displacement curve corresponding

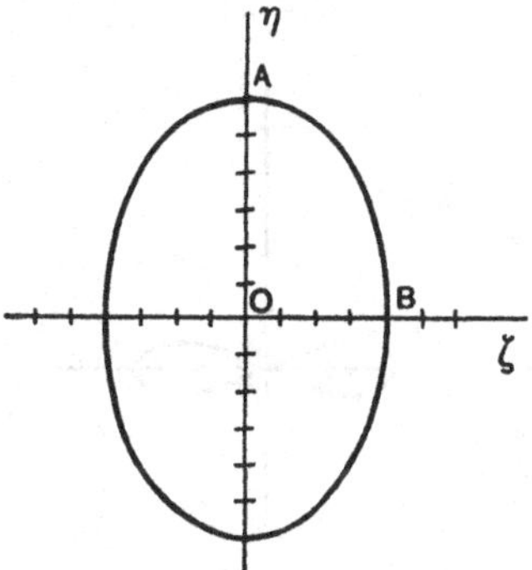

Fig. 1.
$a=4$

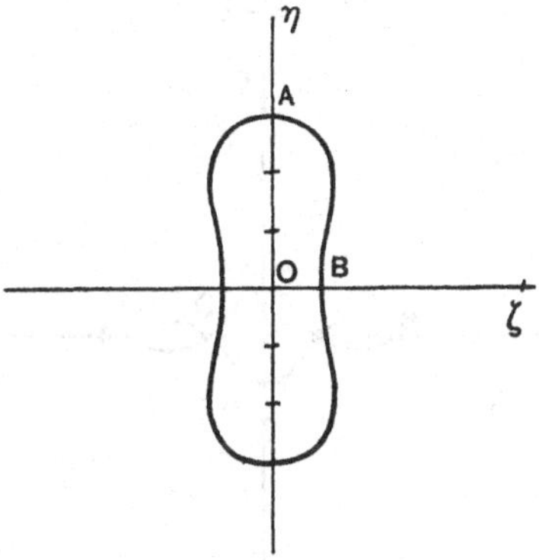

Fig. 2.
$a=1$

* Cf. §§ 11, 13.

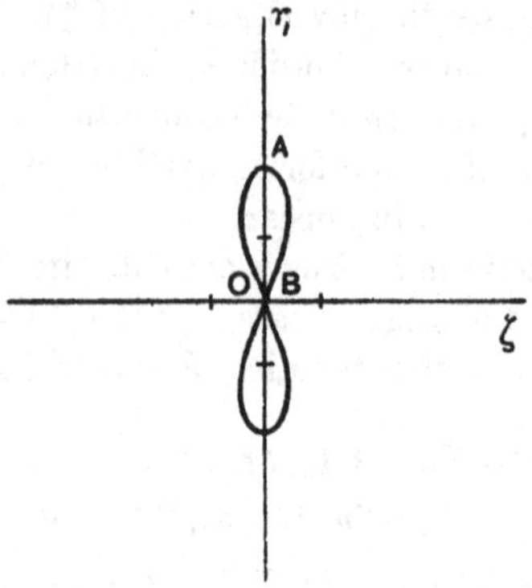

Fig. 3.
$a=0$

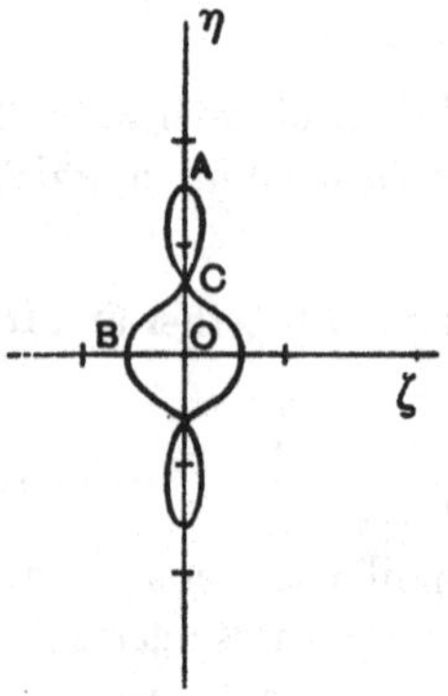

Fig. 4.
$a=-\frac{1}{2}$

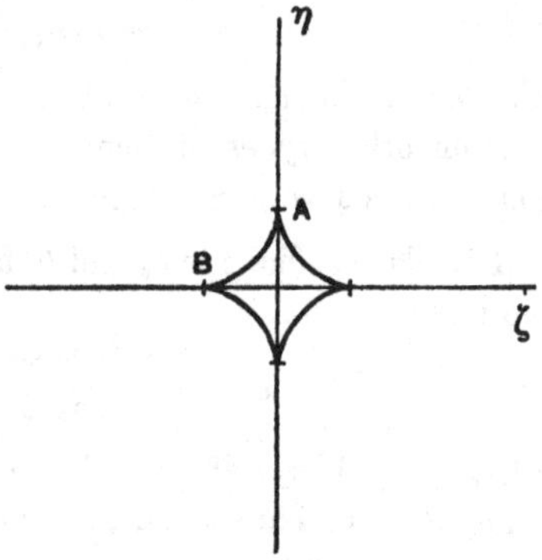

Fig. 5.
$a=-1$

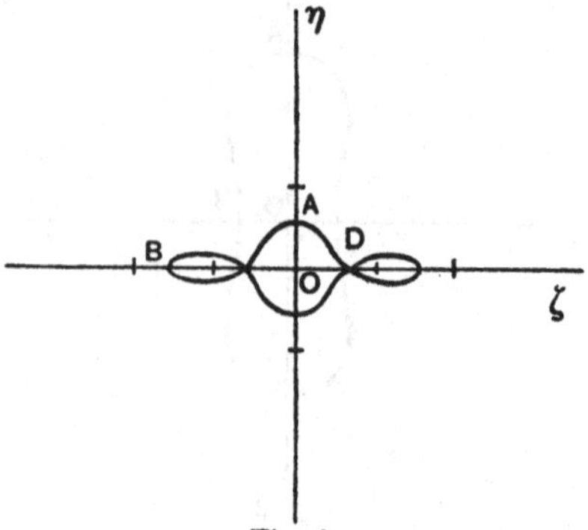

Fig. 6.
$a=-\frac{3}{2}$

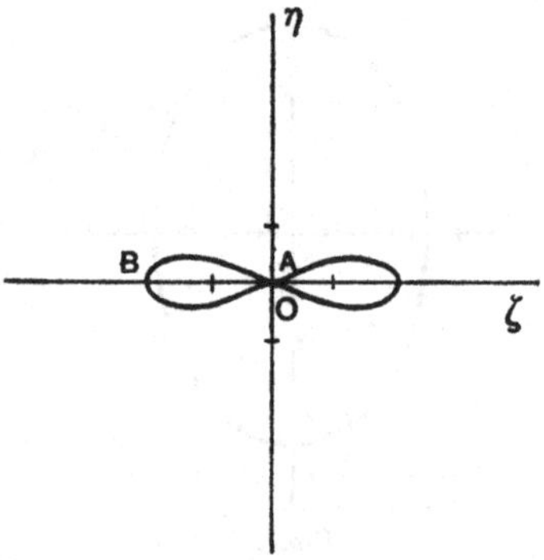

Fig. 7.
$a=-2$

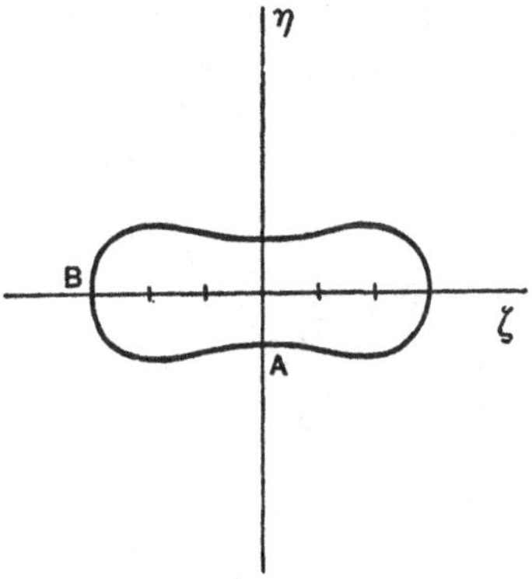

Fig. 8.
$\alpha = -3$

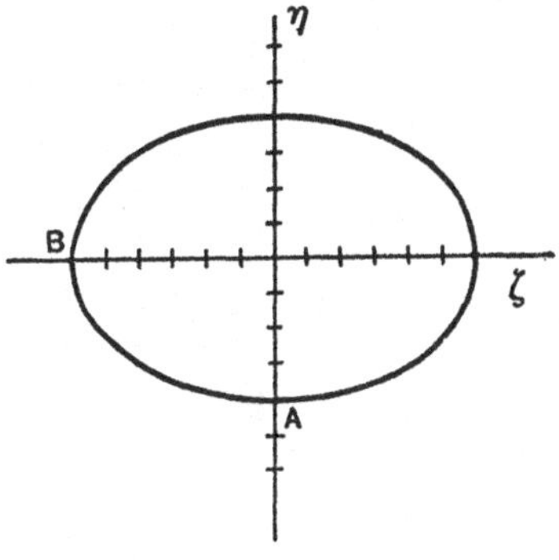

Fig. 9.
$\alpha = -6$

to a given value of ρ, the radius of the exit-pupil. The curve varies considerably in appearance according to the value of α, and the diagrams shew some typical cases. For varying values of ρ, i.e., of the exit-pupil, the curves vary in size, but not in shape, inversely as the cube of ρ; so that each of the curves may be regarded as the shape of the illuminated area for that particular value of α. Fig. 5 is of interest; its equation is

$$\eta^{\frac{2}{3}} + \zeta^{\frac{2}{3}} = 1,$$

and gives the minimum area for a given value of b_8.

11. Consider now the terms in b_5 and b_7 and substitute in (4) § 8; we have

$$d^5(Y' - Y_1) = 2\rho^2 Y_1^3 \{b_5(2\cos^2\phi_1 + 1) + 12b_7\cos^2\phi_1\},$$
$$d^5(Z' - Z_1) = 4\rho^2 Y_1^3 b_5 \cos\phi_1 \sin\phi_1;$$

or, as in the previous paragraph, we may write

$$\eta = (1+\alpha)\cos 2\phi_1 + 2 + \alpha, \quad \zeta = \sin 2\phi_1, \quad \text{...........(1)}$$

where $\alpha b_5 = 6b_7$. In general (1) represents an ellipse with centre C upon the axis of η, and symmetrically placed with respect to that axis; and this corresponds to a narrow rim of the exit-pupil of radius ρ. For varying values of ρ all the ellipses so obtained will touch two fixed straight lines passing through the origin; so that we have something very similar to coma, discussed in § 5, and again we have a double description of the curve for a single description of the exit-pupil. This aberration may be named 'elliptical' coma: the angle between the tangents is $2\cot^{-1}\sqrt{3+2\alpha}$. It is clear that if b_5 vanishes we have just a radial distortion along the η-axis and the aberration curve reduces merely to a portion of this axis; if $\alpha + 1 = 0$ the curve is a straight line perpendicular to the η-axis, for one of the axes of the ellipse will vanish. In this case corresponding to a point source of light in the object space

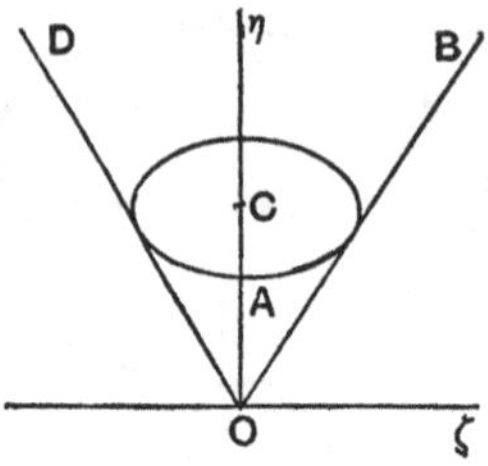

Fig. 1.

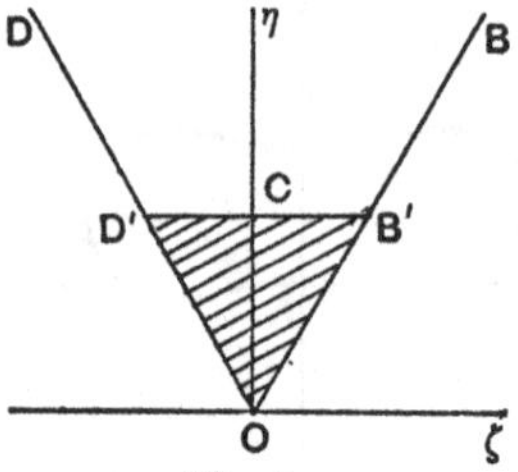

Fig. 2.

we have a triangular patch of illumination upon the conjugate plane, as in Fig. 2.

12. The terms in b_3 and b_4 fall naturally together; but they may be considered in a more general way; for if we have terms $\alpha\theta^{n-1}\psi$ and $\beta\theta^{n-2}\phi^2$ appearing in the aberration function F then upon substitution in (4) § 8 we obtain

$$\left.\begin{aligned} d^{2n-1}(Y'-Y_1) &= 2\rho Y_1^{2n-2}(\alpha+4\beta)\cos\phi_1, \\ d^{2n-1}(Z'-Z_1) &= 2\rho Y_1^{2n-2}\alpha\sin\phi_1. \end{aligned}\right\} \qquad \text{.........(1)}$$

These relations clearly represent astigmatism of order $(n-1)$ and they indicate two images and two image surfaces according as we consider a primary or a secondary plane; and the surfaces coincide only if $\beta=0$ —the field being flat if in addition $\alpha=0$. In general the astigmatic separation δ is given by

$$\delta d^{2n-2} = 8\beta Y_1^{2n-2}. \qquad \text{.........(2)}$$

And again the term in b_2 may be examined from the general point of view; for a term $\alpha\theta^{n-1}\phi$ in the aberration function F will imply, by (4) §8, the relations

$$d^{2n-1}(Y'-Y_1) = 2\alpha Y_1^{2n-1}, \quad Z'-Z_1=0;$$

and this clearly indicates distortion proportional to the $(2n-1)$th power of Y_1, being 'barrel' or 'pin-cushion' distortion according as $\alpha \lesseqgtr 0$.

13. *The Third Order Aberrations.* We may investigate the aberrations of the third order by considering F to be a quartic function of the variables θ, ϕ and ψ; but the aberration curves introduced will be very similar in appearance to those discussed in previous sections—with one exception*. And this exception will arise from the terms in F of the form

$$c_9\theta\phi\psi^2 + c_{12}\phi^3\psi;$$

substituting in (4) § 8 we have corresponding displacements given by

$$d^7(Y' - Y_1) = 2\rho^4 Y_1^3\{8c_{12}\cos^4\phi_1 + (12c_{12} + 4c_9)\cos^2\phi_1 + c_9\},$$
$$d^7(Z' - Z_1) = 2\rho^4 Y_1^3\{8c_{12}\cos^2\phi_1 + 4c_9\}\sin\phi_1\cos\phi_1.$$

These may be written, as in preceding paragraphs,

$$\left.\begin{aligned}\eta &= \cos 4\phi_1 + 2(5 + 4a)\cos 2\phi_1 + 9 + 12a,\\ \zeta &= \sin 4\phi_1 + 2(1 + 4a)\sin 2\phi_1,\end{aligned}\right\}\quad\ldots\ldots\ldots\ldots(1)$$

where $c_9 = 4ac_{12}$, η and ζ being current co-ordinates referred to parallel axes with the Gaussian image as origin. These displacement curves are periodic in ϕ_1, the period being π, so that half the rim of the exit-pupil corresponds to the whole perimeter of the corresponding curve; moreover they are symmetrical about the line $\phi_1 = 0$, i.e., about the η-axis. And for a given value of a, and varying values of ρ, they all touch fixed tangents passing through the origin of co-ordinates; but these tangents become imaginary for certain ranges of values of a. It is clear, therefore, that we are dealing with an aberration of a generalised coma type.

The aberration curves vary in appearance with the value of a; the diagrams given shew a representative selection of such values, and it is to be understood that each curve corresponds to an annulus (or rather to half an annulus) of the exit-pupil of radius ρ—the dimensions of the curve varying as the fourth power of ρ.

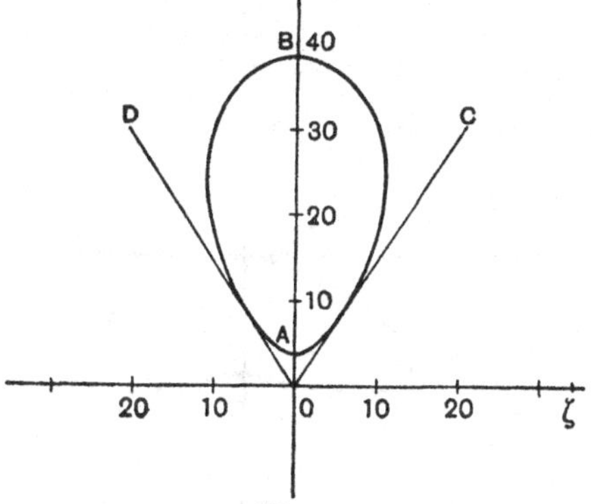

Fig. 1.
$a = 1$

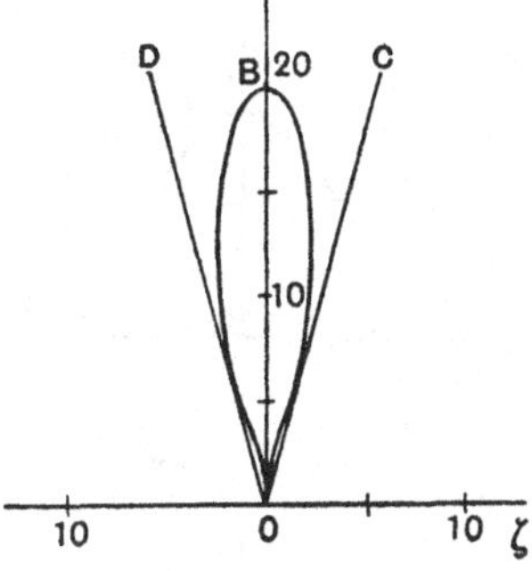

Fig. 2.
$a = 0$

* Cf. below, § 16.

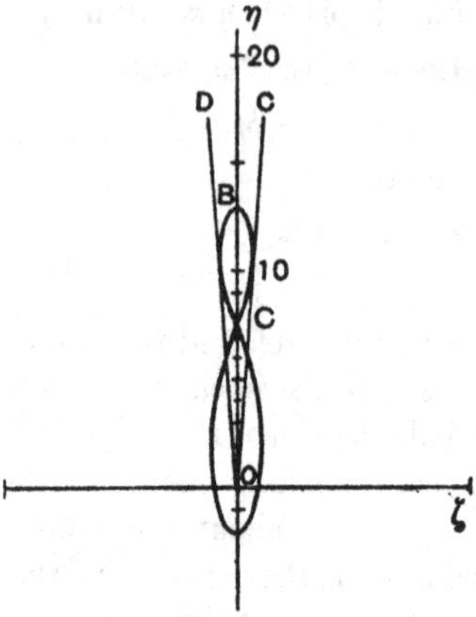

Fig. 3.
$8a+3=0$

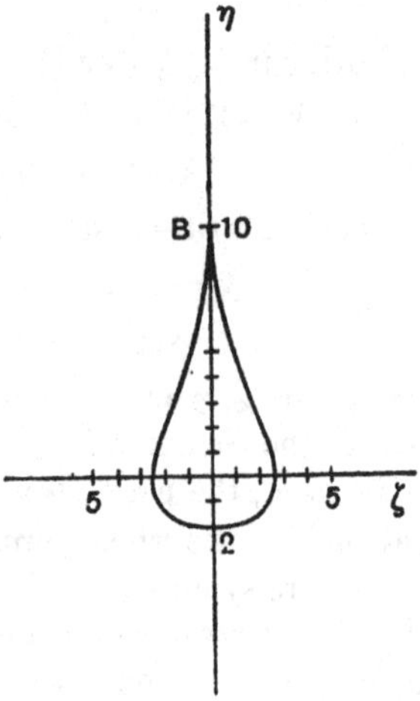

Fig. 4.
$2a+1=0$

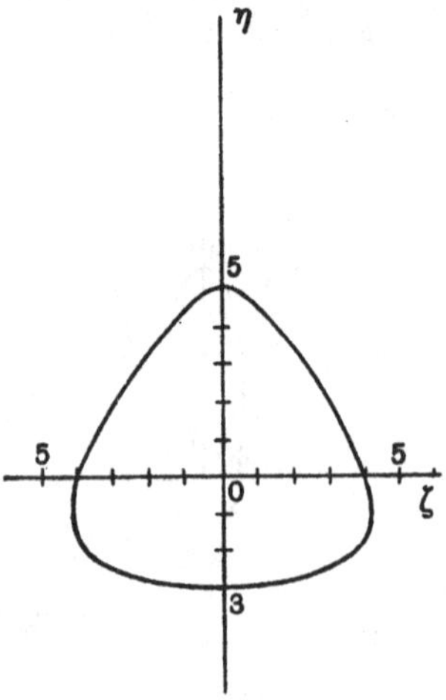

Fig. 5.
$4a+3=0$

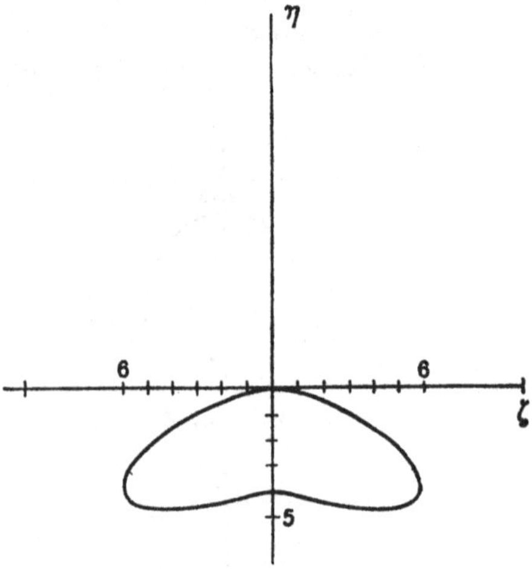

Fig. 6.
$a+1=0$

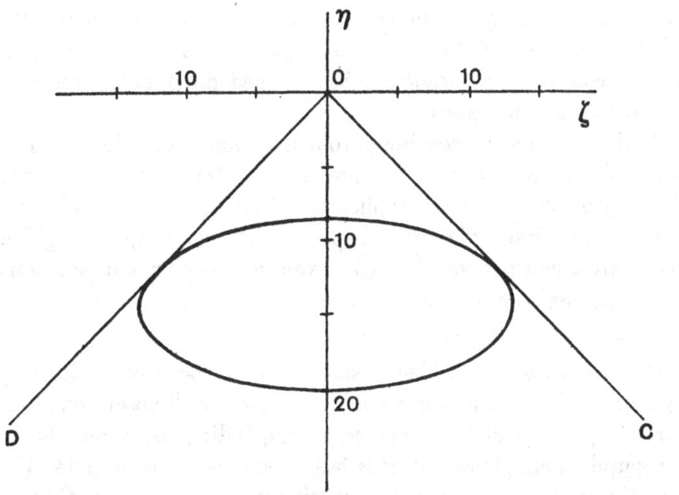

Fig. 7.
$a+2=0$

14. *The Combination of Aberrations**. In the preceding paragraphs the aberrations of any particular order have been considered alone; but in fact the displacements of the second order will be modified by the coefficients of the first order—assuming that these have not been eliminated; so that there is revealed the possibility of the combination of aberrations, and further, of the balancing of aberrations. As a simple example we find that, if we consider only aberrations of the first two orders and confine ourselves further to an object point situate upon the axis of symmetry, the radius of the illuminated disc upon the paraxial image plane is

$$4a_6\rho_1^3 + 6(a_8 + b_{10})\rho_1^5,$$

ρ_1 being the radius of the exit-pupil. The second order displacement will vanish, therefore, for all values of ρ_1 if $a_8 + b_{10} = 0$, while spherical aberration, as a whole, will vanish only if both coefficients are zero; but very frequently a lens is designed so that the paraxial and marginal foci coincide at full aperture (or at a partial aperture) and then we have

$$2a_6 + 3(a_8 + b_{10})\rho_1^2 = 0, \quad \text{..................(1)}$$

a relation giving a_6 when b_{10} is known. So that the effect of axial aberration can be diminished by a proper choice of the *a*-coefficient—supposed under control—without a change of the *b*-coefficient. Higher order axial aberrations often have to be taken into account, as for example in the design of microscope objectives, but the principle remains

* Cf. *Trans. Camb. Phil. Soc.* XXIII, No. 9, §§ 12, 13, 20 and 21.

the same; and again similar effects must be considered with the other aberrations of these and of higher orders. Thus the aberration displacements of any order can be modified profoundly by a choice of the coefficients of lower order.

Again we have to consider, from a geometrical point of view, the effect of change of focus in the presence of aberrations in general; and this may be effected by an application of § 2. In particular with circular coma of the first order, the displacement curves upon neighbouring planes are somewhat complicated—even those corresponding to a narrow rim of the exit-pupil. But this may be contrasted with the diffraction phenomena to be investigated later*.

15. *Change of Focus.* The effect of change of focus may be investigated by means of the results given in § 2, and we shall obtain displacement curves, upon out-of-focus planes, corresponding to an annulus of the exit-pupil; a simple case of this has already been given in § 6. We may investigate also, in this way, the caustic surfaces arising from the presence of aberrations of various types. The out-of-focus aberration curves corresponding to the comatic aberrations will in general be of a complicated type; as a simple example we discuss the case arising from first order circular coma. Thus combining § 2 with § 5 we find that the displacement, upon an out-of-focus plane, due to this aberration is given by

$$\eta = \cos 2\phi_1 - \alpha \cos \phi_1 + 2 + \beta, \quad \zeta = \sin 2\phi_1 - \alpha \sin \phi_1. \quad \ldots\ldots(1)$$

Here $\quad d^3(Y'' - Y_1) = 2a_5 Y_1 \rho^2 \eta, \quad d^3(Z'' - Z_1) = 2a_5 Y_1 \rho^2 \zeta,$

and $\quad X d^2 = 2a_5 Y_1 \rho \alpha = 2a_5 \rho^2 \beta;$

so that η and ζ may be taken as current co-ordinates in the usual way. Equations (1) give the displacement curves corresponding to an annulus of radius ρ of the exit-pupil and the numerical term 2 indicates that the origin has moved along the η-axis a distance proportional to ρ^2.

The curves are clearly periodic in ϕ, the period being 2π here instead of π, as for the Gaussian plane (i.e., $\alpha = 0$); also they are symmetrical about the η-axis. Moreover the curves on one side of the Gaussian plane $\alpha = 0$ are exactly similar to those on the other side except that they are rotated through two right-angles.

Three curves are shewn in the diagram corresponding to the focussing planes $\alpha = \frac{1}{2}$, 1 and 2; for the complete circular aperture the effect will be due to a superposition of these curves combined with a progressive change of origin proportional to the square of the radius of the exit-pupil. This complicated effect may be compared with the modification due to diffraction to be investigated subsequently*.

* Cf. Chapter VII, § 5.

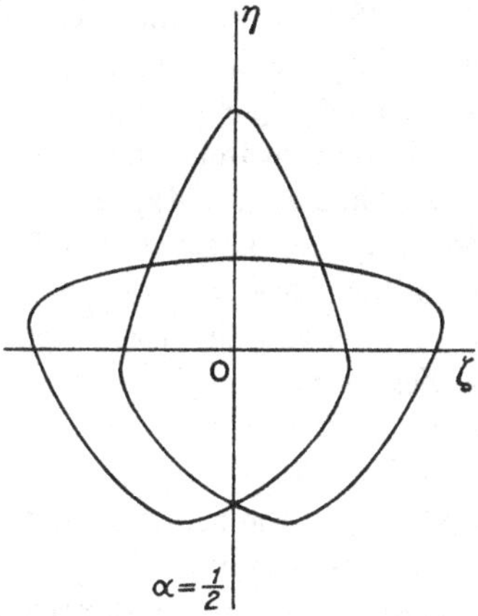

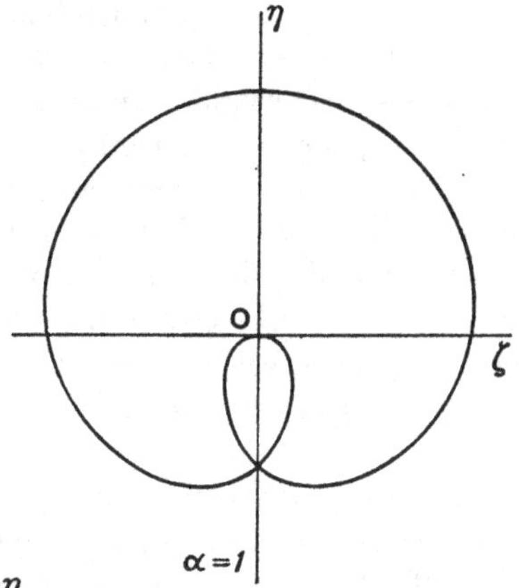

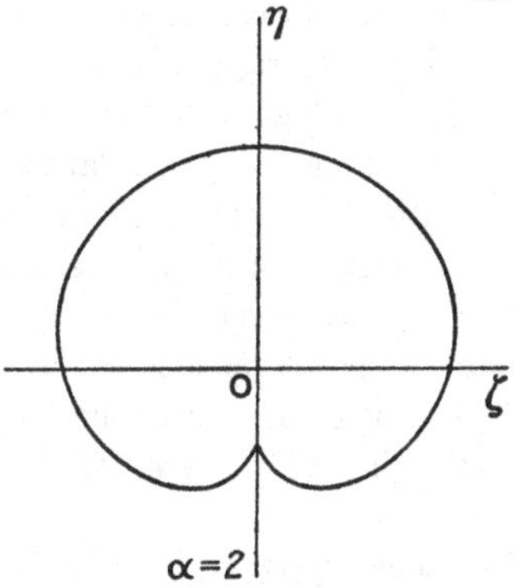

16. *Some General Remarks.* Let us consider the aberrations of any particular order, in the absence of those of lower orders; we have an aberration function F and corresponding displacements as in (4) § 8, and combining these with the formulae of § 2 for out-of-focus effects the whole displacement upon a receiving plane normal to the axis of the system is given by a sum of a number of terms alternatively of the types S and C given below, viz.,

$$\eta = (\sum_{s=0} A_{2s} \cos^{2s} \phi_1 - \mu) \cos \phi_1, \quad \zeta = (\sum_{s=0} B_{2s} \cos^{2s} \phi_1 - \mu) \sin \phi_1, \ldots\ldots S$$

and

$$\eta = \sum_{s=0} A'_{2s} \cos^{2s} \phi_1 - \mu \cos \phi_1, \quad \zeta = (\sum_{s=0} B'_{2s} \cos^{2s} \phi_1) \sin \phi_1 \cos \phi_1 - \mu \sin \phi_1, \ldots C$$

Here η and ζ are current co-ordinates relative to the central point as origin and μ denotes a displacement from the paraxial image plane; in fact $\mu d = X\rho$, in the usual notation. The coefficients contain as factors powers of ρ and Y_1, and those coefficients occurring in any one of the types S and C contain severally some particular power of ρ and a particular power of Y_1; moreover the S-type arises from even powers of q in § 8 so that ρ occurs to an odd power, while in the C-type ρ occurs to an even power. All the aberrations may be classified therefore according as they fall under one or other of the above two types, which may be named respectively the 'spherical' type and the 'coma' type; or the 'symmetrical' and the 'unsymmetrical' type.

For the spherical type let us write, in the first place, $\mu = 0$ and so consider the paraxial image plane; it is evident that the corresponding displacement curves are closed and periodic in ϕ_1, the period being 2π, and that an increase in ϕ_1, of amount π, serves merely to change the signs of both η and ζ; moreover the curves are symmetrical with respect to the co-ordinate axes. Thus the origin is within (or upon) each curve and as ρ decreases from its maximum value to zero each curve collapses to a point at the origin, remaining always similar and similarly situated. This may be compared with the simplest case of this type, viz., axial spherical aberration. Again change of focus is beneficial, in general, i.e., the area of the resulting light patch may be diminished in this way, and we may find the least curve of aberration—corresponding to the least circle of aberration of § 4. It is evident that the largest value of ρ, the radius of the exit-pupil, will give the area of illumination upon the receiving plane.

Turning now to the coma type, it is evident that, if $\mu = 0$, the summations C may be written as functions of $2\phi_1$, the equation for η involving also a term independent of ϕ_1 but containing ρ and Y_1; and this latter term implies a change of origin along the η-axis. The displacement curves are symmetrical with respect to the axis of η and are periodic in ϕ_1, their period being π, so that half of the perimeter of the exit-pupil corresponds to the whole of the aberration curve. For varying values of ρ the new origin moves along the η-axis and two cases arise; either (1) any curve, and so each curve, has the old origin, i.e., the Gaussian image, within (or upon) it; or (2) the origin is outside each curve. In the second case we have a family of curves all touching real fixed tangents passing through the Gaussian image and collapsing to a point there as ρ tends to zero; and this may be compared with circular coma discussed in § 5. In the former case we have something similar, in some respects, to the spherical type of aberration; the fixed tangents have now become imaginary.

Again, in the more general case in which $\mu \neq 0$, and we consider an out-of-focus plane, it is clear from the summations above that an addition of π to ϕ_1, together with a change in the sign of μ, leave the expressions C unaltered; so that the displacement curves are symmetrical with respect to the paraxial image plane, indicating that change of focus is of no advantage with the coma type of aberration.

Each order of aberration brings with it an aberration of a new kind, for an additional power of $\cos^2 \phi_1$ appears in one or other of the summations S and C; and further the new kind of aberration falls alternately under these types. If the order of the aberration be even the new displacement is of the spherical type, if odd it is of the coma type: thus the only aberration of zero order is that due to a shift in the focussing plane and is spherical in type; the essentially new aberration of order one is coma, of order two it is that examined in § 10 and belongs to the spherical type, while of order three it is as in § 13 and is comatic.

REFERENCES

Pelletan, A. *Optique Appliquée.*

Rayleigh. Hamilton's Principle and the Five Aberrations of von Speidal. *Sci. Papers*, vol. v, p. 456.

Steward, G. C. *Trans. Camb. Phil. Soc.* XXIII, No. 9 (1926).

—— *Phil. Trans. Roy. Soc.* A, 225, 1925, Part I.

—— 'On the Optical Writings of Sir William Rowan Hamilton,' *Math. Gazette*, July 1932, vol. XVI, no. 219.

von Seidel. *Astron. Nachrichten*, 835, 871, 1027, 1028, 1029.

CHAPTER IV

THE SINE-CONDITION AND THE OPTICAL COSINE LAW

1. *The Sine-Condition.* Object glasses of telescopes, after being polished, are usually finished off by a process of 'figuring'; i.e., the material of the glass is removed so that a non-spherical surface of revolution is obtained and the process is continued until the definition for points near to the axis of the system is found to be satisfactory. The method is therefore one of trial and error. Now very good object glasses were fashioned by Fraunhofer in this way and it was found long afterwards that they satisfied very nearly a certain condition—the well-known 'sine-condition'; and the better the definition near the axis the more nearly did they satisfy the condition. It is our object to investigate the reason for this and to find out the meaning of this 'sine-condition.'

The meaning of 'aplanatism' has been considered in Chapter I; and defined there as the satisfaction of *two* conditions, one implying the absence of spherical aberration and the other the satisfaction of a certain relation between the sines of the inclinations of the two parts of a ray to the axis of the system. We consider now the general symmetrical system.

Let θ and θ' be the inclinations to the axis of the two parts of a ray before and after refraction; then in the notation of § 10, Chapter II, we have

$$\left(\frac{\sin\theta}{\sin\theta'}\right)^2=\frac{M^2+N^2}{M'^2+N'^2}=m^2\frac{\left(\frac{\partial V}{\partial Y_1}\right)^2+\left(\frac{\partial V}{\partial Z_1}\right)^2}{\left(\frac{\partial V}{\partial y'}\right)^2+\left(\frac{\partial V}{\partial z'}\right)^2},$$

i.e.,
$$\left(\frac{\sin\theta}{m\sin\theta'}\right)^2=\frac{\theta\left(\frac{\partial V}{\partial\theta}\right)^2+\phi\frac{\partial V}{\partial\theta}\frac{\partial V}{\partial\phi}+\psi\left(\frac{\partial V}{\partial\phi}\right)^2}{\theta\left(\frac{\partial V}{\partial\phi}\right)^2+\phi\frac{\partial V}{\partial\phi}\frac{\partial V}{\partial\psi}+\psi\left(\frac{\partial V}{\partial\psi}\right)^2},$$

θ, ϕ and ψ being our usual variables and L, M, N and L', M', N' the direction cosines of the incident and emergent rays. If now we assume an axial object point, i.e., $Y_1=mY=0$ and $Z_1=mZ=0$, we shall have $\theta=\phi=0$, while as usual $\psi d^2=\rho^2$; so that

$$\left(\frac{\sin\theta}{m\sin\theta'}\right)^2=\left(\frac{\partial V}{\partial\phi}\Big/\frac{\partial V}{\partial\psi}\right)^2.$$

Write now
$$V=f(\theta)-d\sqrt{1+u}+F,$$
where F is the aberration function; then

$$\frac{\partial V}{\partial\phi}=\frac{d}{2\sqrt{1+u}}+\frac{\partial F}{\partial\phi},$$

and
$$\frac{\partial V}{\partial\psi}=-\frac{d}{2\sqrt{1+u}}+\frac{\partial F}{\partial\psi},$$

if $u=\theta-\phi+\psi$; so that

$$\left(\frac{\sin\theta}{m\sin\theta'}\right)^2=\left(\frac{1+\frac{2\sqrt{1+u}}{d}\frac{\partial F}{\partial\phi}}{1-\frac{2\sqrt{1+u}}{d}\frac{\partial F}{\partial\psi}}\right)^2. \quad\text{.........(1)}$$

Now F is a function of θ, ϕ and ψ and we may write

$$F=\Sigma\Sigma\Sigma A_{p,q,r}\,\theta^p\phi^q\psi^r,$$

where $p+q+r\geqslant 2$: the A's are of course the aberration coefficients whose meanings have been investigated in the preceding chapter. Since $\theta = \phi = 0$, we have

$$\frac{\partial F}{\partial \phi} = \sum_{n=1}^{\infty} A_{0,1,n} \psi^n, \quad \text{......................(2)}$$

and

$$\frac{\partial F}{\partial \psi} = \sum_{n=2}^{\infty} n A_{0,0,n} \psi^{n-1}; \quad \text{...................(3)}$$

and the coefficients of (3), viz., $A_{0,0,n}$, are those which we have shewn to govern pure axial spherical aberration of the several orders; in the absence, therefore, of axial spherical aberration

$$\frac{\partial F}{\partial \psi} = 0,$$

and

$$\left(\frac{\sin \theta}{m \sin \theta'}\right)^2 = \left\{1 + \frac{2}{d}\sqrt{1 + \frac{\rho^2}{d^2}} \sum_{n=1}^{\infty} A_{0,1,n} \psi^n\right\}^2; \quad \text{.........(4)}$$

and if the sine-condition be satisfied for all rays, i.e., if

$$\sin \theta = m \sin \theta', \quad \text{..........................(5)}$$

then

$$\frac{2}{d}\sqrt{1 + \frac{\rho^2}{d^2}} \sum_{n=1}^{\infty} A_{0,1,n} \left(\frac{\rho}{d}\right)^{2n} = 0$$

for all values of ρ.

It follows therefore by taking successive terms that each of the coefficients $A_{0,1,n}$ must vanish; conversely if these vanish the sine-condition (5) is satisfied for all rays through the conjugate axial points considered. Now we have seen in the previous chapters that the coefficients $A_{0,1,n}$ are those which govern circular coma of the several orders—which depends upon the first power of the quantity (Y_1/d); and this, in the absence of axial aberration, is the most important departure from perfect imagery for points near to the axis of the system. The satisfaction of the sine-condition, therefore, in the absence of spherical aberration will ensure a good image of a small two-dimensional *object* placed normal to the axis, for it implies the absence of circular coma of *all* orders for the conjugate planes under consideration.

2. It is of interest to investigate the geometrical meaning of a given departure from the sine-condition; thus let the paraxial magnification be m for the conjugate points considered and let the magnification as calculated from the sine-condition applied to a marginal ray be $m(1+\epsilon)$; it is assumed that spherical aberration is absent for these conjugate points. Then from (4) § 1 we have

$$\epsilon = \frac{2}{d}\sqrt{1 + \left(\frac{\rho_1}{d}\right)^2} \sum_{n=1}^{\infty} A_{0,1,n} \left(\frac{\rho_1}{d}\right)^n, \quad \text{..............(1)}$$

and $A_{0,1,n}$ are the coefficients governing circular coma of the several orders; in the notation of § 5, Chapter III we have

$$d^{2n+1} P_1 G_n = 2A_{0,1,n} Y_1 \rho^{2n},$$

the suffix indicating that we are dealing with circular coma of order n. From (1) therefore

$$\epsilon Y_1 = \sqrt{1+\left(\frac{\rho_1}{d}\right)^2} \sum_{n=1}^{\infty} P_1 G_n, \quad \text{.................(2)}$$

and here ρ_1 is the radius of the exit-pupil. If we consider only the first order aberration

$$\epsilon Y_1 = PG_1,$$

a result of some importance subsequently.

For a ray cutting the exit-pupil at a distance ρ from the axis the corresponding value of ϵ is given by

$$\epsilon Y_1 = \sqrt{1+\left(\frac{\rho}{d}\right)^2} \sum_{n=1}^{\infty} \left(\frac{\rho}{\rho_1}\right)^{2n} P_1 G_n,$$

where now $P_1 G_n$ corresponds to ρ.

3. *The Cosine-Conditions.* We have seen already that for all rays through a given object point

$$\left.\begin{array}{rl} & M - mM' = \text{const.} \\ \text{and} & N - mN' = \text{const.} = 0 \end{array}\right\} \quad \text{.....................(1)}$$

if $Z_1 = mZ = 0$; and this is true only in an aberrationless system.

This result is an extension of the well-known sine-condition into the outer parts of the field and (1) may be named the 'cosine-conditions.' Departures from this result will imply the presence of aberrations and we have here a condition applying to the higher order aberrations in the same way as the ordinary sine-condition applies to circular coma.

4. *Herschel's Condition.* Suppose that we have two conjugate axial points P and P' which are free from spherical aberration of all orders; in general neighbouring and conjugate axial points Q and Q' will not be free from aberration but we may enquire what condition must be satisfied so as to ensure that Q and Q' shall be free from aberration. Herschel gave this condition in an approximate form, applying only to first order aberration, i.e., to aberration depending upon the cube of the inclination of the ray to the axis; and later Abbe shewed that the condition was contained in a simpler—and more general—result; the final form of the condition is

$$\sin \theta/2 = m \sin \theta'/2. \quad \text{.......................(1)}$$

Here θ and θ' are the angles of inclination of the ray to the axis of the system and m is the reduced magnification associated with the conjugate points considered. The condition is customarily written

$$\mu \sin \theta/2 = \mu' m \sin \theta'/2, \qquad \text{.....................(2)}$$

but here m is the ordinary magnification.

In the usual notation we have

$$V = f(\theta) - d\sqrt{1+u} + F;$$

and from the general property of the characteristic function

$$L = -\frac{\partial V}{\partial x} \text{ and } L' = \frac{\partial V}{\partial x'}, \qquad \text{.....................(3)}$$

all distances being reduced. But $x = 1/m$ and $x' = -s$, where s is the paraxial magnification associated with the pupil planes; so that

$$L = m^2 \frac{\partial V}{\partial m} \text{ and } L' = -\frac{\partial V}{\partial s}. \qquad \text{..................(4)}$$

We are dealing with axial conjugates so that we shall have $\theta = \phi = 0$; moreover

$$\theta d^2 = Y_1^2 + Z_1^2, \quad \phi d^2 = 2(Y_1 y' + Z_1 z') \text{ and } \psi d^2 = y'^2 + z'^2 = \rho^2,$$

so that

$$\frac{\partial \psi}{\partial m} d - 2\psi = 0 \text{ and } \frac{\partial \psi}{\partial s} d + 2\psi = 0.$$

If we write
$$F = \sum_{n=2}^{\infty} A_n \psi^n$$

in (4) we have

$$\left.\begin{aligned} L &= m^2 \frac{\partial f(0)}{\partial m} + \frac{m^2}{\sqrt{1+\psi}} + m^2 \sum_{n=2}^{\infty} \psi^n \left(\frac{2n}{d} A_n + \frac{\partial A_n}{\partial m}\right) \\ \text{and} \qquad L' &= \frac{1}{\sqrt{1+\psi}} + \sum_{n=2}^{\infty} \frac{2n}{d} A_n \psi^n \end{aligned}\right\} \text{......(5)}$$

since A_n is independent of s, as also is $f(0)$; indeed if F_1 and F_2 be the principal foci of the system and P and P' the axial conjugates associated with reduced magnification m we have

$$F_1 P = 1/m \text{ and } F_2 P' = -m,$$

so that
$$\begin{aligned} f(0) &= PP' = F_1 F_2 - F_1 P + F_2 P' \\ &= F_1 F_2{}^* - 1/m - m; \end{aligned}$$

and
$$\frac{\partial f(0)}{\partial m} = \frac{1}{m^2} - 1.$$

* The modified and reduced axial distance between the principal foci of the system.

Thus from (5) above

$$\frac{L-1}{L'-1}=m^2\frac{1-1/\sqrt{1+\psi}-\Sigma-\Sigma'}{1-1/\sqrt{1+\psi}-\Sigma}, \quad \dots\dots\dots\dots\dots(6)$$

where $$\Sigma\equiv\sum_{n=2}^{\infty}\frac{2n}{d}A_n\psi^n \text{ and } \Sigma'\equiv\sum_{n=2}^{\infty}\psi^n\frac{\partial A_n}{\partial m}.$$

If now the conjugate axial points P and P' be free from spherical aberration, $A_n=0$ for all values of n and thus $\Sigma=0$; so that from (6), θ and θ' being the initial and final inclinations of the ray to the axis of the system,

$$\frac{\sin\theta/2}{\sin\theta'/2}=m\left\{1-\frac{\Sigma'}{1-(1+\psi)^{-\frac{1}{2}}}\right\}^{\frac{1}{2}}. \quad \dots\dots\dots\dots\dots(7)$$

It is clear therefore that if $\frac{\partial A_n}{\partial m}=0$ for all values of n, i.e., if $\Sigma'=0$, then the condition (1) is satisfied for all rays; and conversely by the expansion of (7) it follows that if the condition (1) be satisfied for all rays then $\frac{\partial A_n}{\partial m}=0$ for all values of n.

Thus Herschel's condition, applied to two conjugate axial points free from spherical aberration, implies that neighbouring axial points are free from spherical aberration of *all* orders; and conversely.

In order to investigate the meaning of a departure from Herschel's condition let us write

$$\left(\frac{1}{m}\frac{\sin\theta/2}{\sin\theta'/2}\right)^2=1-\epsilon$$

for any ray; then by comparison with (5) we have

$$\epsilon=\frac{\sum_{n=2}^{\infty}A_n'\psi^n}{1-1/\sqrt{1+\psi}}=(1+\psi+\sqrt{1+\psi})\left(\sum_{n=2}^{\infty}A_n'\psi^n\right),$$

where the dash denotes differentiation with respect to m; so that

$$\epsilon=2A_2'\psi^2+(\tfrac{3}{2}A_2'+2A_3')\psi^3+\dots. \quad \dots\dots\dots\dots\dots(8)$$

Considering therefore only a first approximation, if the magnification as calculated from Herschel's condition applied to a marginal ray be $m(1-\eta)$, we have

$$\frac{\partial A_2}{\partial m}=\eta d^2/\rho_1^2, \quad \dots\dots\dots\dots\dots\dots\dots\dots\dots(9)$$

ρ_1 being the radius of the exit-pupil; and this gives a measure of the spherical aberration introduced by a slight axial movement of the object plane.

It is seen that Herschel's condition (1) is, in general, incompatible with the sine-condition; the only case in which both conditions may be satisfied is that in which $m^2 = 1$, i.e., $\theta' = \pm\,\theta$. Thus it is impossible, in general, to reproduce, without aberration, an element of space axially situated.

5. *Geometrical Proof of the Sine-Condition.* An interesting proof of the sine-condition, and also of Herschel's condition, was given by Hockin in 1884, but it is approximate only and deals with small quantities of the first order. It depends upon the fact that, if there be perfect correspondence between two points, the one in the object space and the other in the image space, then the optical path between them is constant. A slight modification of Hockin's proof is given here.

Let P and P' be two conjugate points upon the axis of a symmetrical optical system and supposed free from aberration and let Q and Q' be

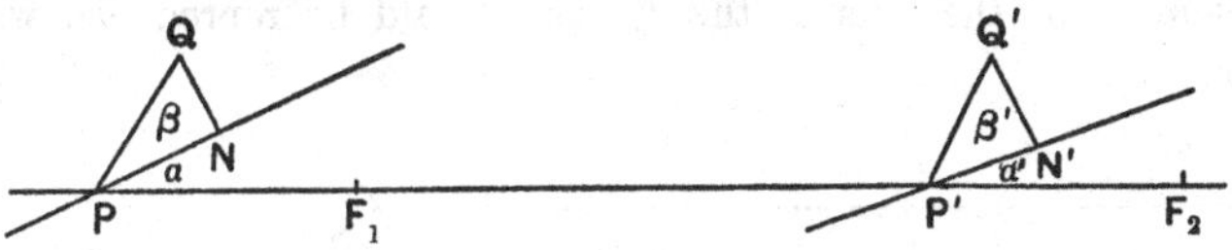

any pair of conjugate points in the neighbourhood of P and P'; we suppose also that Q and Q' are entirely free from aberration. Let PN and $P'N'$ be the corresponding parts of a ray through P and P', making angles α and α' with the axis of the system, and let QN and $Q'N'$ be perpendiculars to this ray. The difference between the optical paths between Q and Q', and P and P', i.e., $V(QQ')$ and $V(PP')$, is given approximately by

$$V(QQ') - V(PP') = \mu' P'Q' \cos\beta' - \mu PQ \cos\beta, \quad \ldots\ldots(1)$$

where $N\hat{P}Q = \beta$ and $N'\hat{P}'Q' = \beta'$; and this must be an absolute constant.

We may make specialisations of this result; thus let PQ, and therefore also $P'Q'$, be normal to the axis PP', i.e., let $\beta = \pi/2 - \alpha$ and $\beta' = \pi/2 - \alpha'$. Then from (1)

$$\mu' P'Q' \sin\alpha' - \mu PQ \sin\alpha = \text{const.}$$
$$= 0, \quad \ldots\ldots\ldots\ldots\ldots\ldots\ldots(2)$$

since the axis PP' is a ray, i.e., $\alpha = 0$ and $\alpha' = 0$ must satisfy the condition. We have, therefore, the usual form of the sine-condition; and this is the condition that there should be perfect correspondence in the neighbourhood of P and P' over normal planes passing through P and P', i.e., we have a good image of a *two*-dimensional object normal to the axis of the system.

Again in (1) let us write $\alpha+\beta=0$ and therefore $\alpha'+\beta'=0$, i.e., let Q and Q' be conjugate axial points; then

$$\mu' P'Q' \cos\alpha' - \mu PQ \cos\alpha = \text{const.}$$
$$= \mu' P'Q' - \mu PQ,$$

since again the condition must be satisfied by $\alpha=0$ and $\alpha'=0$. We have therefore

$$\mu' P'Q' \sin^2 \alpha'/2 = \mu PQ \sin^2 \alpha/2;$$

or if m be the magnification of normal elements, associated with P and P',

$$\mu' m \sin \alpha'/2 = \mu \sin \alpha/2, \quad \text{......................(3)}$$

from the elongation formula of § (3), Chapter I, combined with § (3), Chapter II; here m is the ordinary, not the reduced, magnification. We obtain in this way Herschel's condition—the condition that conjugate and neighbouring axial points should be free from aberration, i.e., that an element of the axis of the system should be reproduced without aberration.

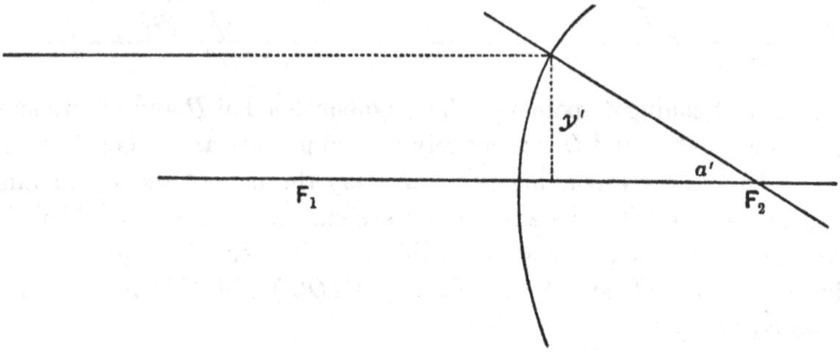

Let us suppose now that the point P moves to an infinite distance in the direction $P'P$ and let F_1 be the first principal focus; P' approaches the second principal focus F_2; using reduced and modified lengths we have

$$F_1 P = 1/m \quad \text{and} \quad \alpha \to 0,$$

so that the sine-condition becomes

$$F_1 P \sin\alpha = \sin\alpha',$$

while the incident ray becomes parallel to the axis: thus if the incident and emergent rays intersect in a point distant y' from the axis we have

$$y' = \sin\alpha',$$

i.e., the locus of this point is a sphere with centre F_2, the second principal focus. The second unit surface is no longer a plane but is a sphere. Other proofs of the sine-condition have been given by Abbe, Helmholtz, and, before these, by Clausius; the latter deduced the condition from the second law of thermodynamics.

6. *The Optical Cosine Law.* The two conditions investigated in the preceding paragraphs, the sine-condition and Herschel's condition, are particular cases of a very general theorem due to T. Smith—the 'optical cosine law'; and this includes also other optical laws. We proceed to prove a lemma leading to this law.

Let $P(x, y, z)$ be a point upon a surface $F(x, y, z)=0$ and let l, m, n be the direction cosines of some *fixed* direction; if L, M, N be the direction cosines of the normal at P to the surface we have

$$\lambda L=\frac{\partial F}{\partial x},\quad \lambda M=\frac{\partial F}{\partial y},\quad \lambda N=\frac{\partial F}{\partial z},\quad \text{where } \lambda^2=\left(\frac{\partial F}{\partial x}\right)^2+\left(\frac{\partial F}{\partial y}\right)^2+\left(\frac{\partial F}{\partial z}\right)^2.$$

Let $Q\ (\xi, \eta, \zeta)$ be a point upon the normal at P to the surface and such that

$$PQ=\rho\cos\theta,$$

ρ being a small constant quantity and θ the inclination of the normal to the direction l, m, n; so that

$$\cos\theta=lL+mM+nN.$$

The equation of the locus of Q is therefore

$$F(\xi-\rho L\cos\theta,\ \eta\mp\rho M\cos\theta,\ \zeta-\rho N\cos\theta)=0,$$

i.e., to the first approximation it is

$$F(\xi, \eta, \zeta)-\rho\cos\theta\left(L\frac{\partial F}{\partial x}+M\frac{\partial F}{\partial y}+N\frac{\partial F}{\partial z}\right)_{\xi, \eta, \zeta}=0. \quad \ldots(1)$$

Again, if $Q'\ (\xi', \eta', \zeta')$ be a point distant ρ' from P in the direction l, m, n, ρ' being small, the equation of the locus of Q' is

$$F(\xi'-\rho' l,\ \eta'-\rho' m,\ \zeta'-\rho' n)=0,$$

i.e., to the first approximation it is

$$F(\xi', \eta', \zeta')-\rho'\left(l\frac{\partial F}{\partial x}+m\frac{\partial F}{\partial y}+n\frac{\partial F}{\partial z}\right)_{\xi', \eta', \zeta'}=0. \quad \ldots\ldots\ldots(2)$$

Equations (1) and (2) will represent the same surface if

$$\lambda\rho\cos\theta=\lambda\rho'\cos\theta,$$

i.e., if

$$\rho=\rho'.$$

Thus the locus of Q is the surface F displaced bodily, without rotation and without distortion, through a small distance ρ parallel to the fixed direction l, m, n.

7. Let F be an incident wave surface in a medium of index μ and let F' be any position of the corresponding emergent wave in a medium of index μ'; let P and P' be corresponding points upon these two surfaces respectively, the reduced path between them being C_1, so that the normals at P and P' are corresponding 'rays.' Suppose the surface F

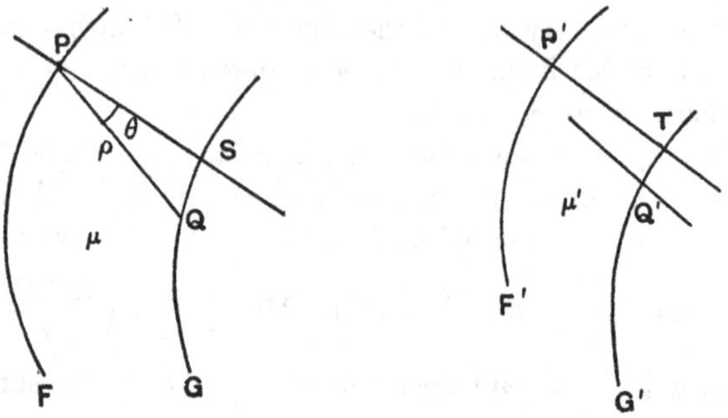

moved, without rotation or distortion, to G through a small distance ρ parallel to some fixed direction inclined at an angle θ to the normal at P, Q being the new position of P; and let G' be a position of the emergent wave surface corresponding to G, the reduced path between them being C_2*; let Q' be the point upon G' corresponding to Q. Then, if the normal at P' intersect G' in T, and the normal at P intersect G in S, we have

$$V(QQ') = V(ST)$$

$$= V(PP') - \mu\rho\cos\theta + \mu' P'T$$

i.e.,
$$\mu' P'T = \mu\rho\cos\theta + V(QQ') - V(PP'),$$

or
$$\mu' P'T = \mu\rho\cos\theta + C_2 - C_1. \quad \text{.................(1)}$$

Suppose now that for all rays we have

$$\cos\theta = \alpha\cos\theta' + \beta, \quad \text{......................(2)}$$

α and β being constants and θ' the inclination of the emergent ray $P'T$ (or normal to the emergent wave surface) to some *fixed* direction in the image space; we may compare (2) with (1), which may be written

$$\mu\rho\cos\theta = \mu' P'T + C_1 - C_2,$$

and since C_2 is arbitrary we may write

$$C_1 - C_2 = \beta\mu\rho;$$

* It is assumed that ρ and $C_2 - C_1$ are small compared with the curvatures of the wave surfaces.

(2) is satisfied for all rays and we have

$$\mu' P'T = \alpha\mu\rho \cos\theta'.$$

By the lemma of the preceding paragraph therefore the locus of T, i.e., the surface G', is merely a bodily displacement of the surface F', without rotation or distortion, through a small distance ρ' parallel to the fixed direction in the image space; and

$$\mu'\rho' = \alpha\mu\rho.$$

And inasmuch as a wave surface implies a corresponding caustic surface we have the result:

If two systems of corresponding rays, touching two caustic surfaces, satisfy the condition $\cos\theta = \alpha\cos\theta' + \beta$, θ *and* θ' *being the inclinations of a ray to two fixed directions, the one in the object space and the other in the image space, and* α *and* β *being constants, then corresponding to a small displacement of either caustic surface, without rotation or distortion, parallel to the fixed direction associated with it, there will be a small displacement, without rotation or distortion, of the corresponding caustic surface parallel to the other fixed direction; and, if the displacements be respectively* ρ *and* ρ' *in media of indices* μ *and* μ', $\mu'\rho' = \alpha\mu\rho$.

This result constitutes the 'optical cosine law'; and it is to be noticed that the fixed directions are any whatever, and not necessarily optically corresponding ones. Moreover the general caustic surface is considered so that the law contemplates the presence of any combination of the geometrical aberrations; and, further, it is not assumed that the optical system is in any way symmetrical, for the only principle employed is that of the 'reduced path.'

8. The optical cosine law gives as special cases a number of optical results and we proceed to consider some of these.

(1) *The Law of Refraction.* Suppose that the optical system consist of a single refracting surface and let the object caustic degenerate to a point upon this surface; then the image caustic will be the same point and the two caustics may be made to undergo the same displacement, viz., along the refracting surface. The cosine law gives for this case

$$\mu' \sin\theta' = \mu \sin\theta + \text{const.},$$

where θ and θ' are the inclinations of a ray with the normal to the refracting surface; and the constant must vanish since the two rays, incident and emergent, may coincide, without deviation, with the normal. We derive therefore the ordinary law of refraction; the law of reflection follows in a similar way.

(2) *The Sine-Condition.* In a system symmetrical about an axis take two axial and conjugate points, supposed free from aberration; the two caustic surfaces reduce to these two points. Then if a slight motion of one caustic (or point) perpendicular to the axis of the system implies a motion of the other caustic—still remaining a point—also perpendicular to the axis the system will be free from the first extra-axial aberration, i.e., will be free from circular coma of all orders. And the condition for this, from the cosine law, is

$$\mu' m \sin\theta' = \mu \sin\theta + \text{const.},$$

θ and θ' being the inclination to the axis of a ray, before and after refraction, and m the magnification; again the constant must vanish since θ and θ' vanish together and we have the sine-condition of § 1.

(3) *Herschel's Condition.* In (2) displacements perpendicular to the axis of the system were considered; take now an exactly similar case but with small displacements along the axes. Then the object caustic being a point the image caustic will remain a point, i.e., no aberrations will be introduced provided that

$$\cos\theta' = \alpha\cos\theta + \text{const.},$$

the ratio of the reduced displacements being α, so that by Maxwell's elongation formula $\alpha = 1/m^2$; and also $\theta = 0$ and $\theta' = 0$ correspond, so that

$$m^2(1 - \cos\theta') = 1 - \cos\theta;$$

whence
$$m \sin\theta'/2 = \sin\theta/2,$$

and we have Herschel's condition that the spherical aberration, supposed zero for two conjugate axial points, should be stationary for small axial displacements of these points.

(4) *The Cosine Relations.* If there be perfect definition for the conjugate planes at magnification m the object and image point caustics must satisfy the cosine law for displacements parallel to the axes of y and z, and of y' and z' respectively, these axes being supposed parallel as usual. The conditions therefore are

$$\mu M - m\mu' M' = \text{const., and } \mu N - m\mu' N' = \text{const.},$$

L, M, N and L', M', N' being the direction cosines of the ray in the two spaces. The constants, of course, will be functions of the co-ordinates but not of the direction cosines; we have, in fact, the cosine elations of § 3.

It will be observed that the cosine law is a generalisation of the sine-condition and of Herschel's condition and that, unlike these conditions, it is valid even in the presence of spherical aberration.

REFERENCES

Abbe. *Archiv für mikroskopische Anatomie*, IX, p. 40, 1873.
Clausius. *Mechanische Wärmetheorie.*
Helmholtz. *Pogg. Annalen*, Jubelband, 1874, p. 557.
Hockin. *Journal of the Royal Microscopical Society*, IV, p. 337, 1884.
Smith, T. The Optical Cosine Law. *Trans. Opt. Soc.* XXIV (1922–3), No. 1.
—— *Dict. App. Physics*, Art. 'Optical Calculations.'
Steward, G. C. *Phil. Trans.* A, 225, Part II, § 21.
—— *Trans. Camb. Phil. Soc.* XXIII, No. 9, § 10.
—— *Proc. Camb. Phil. Soc.* XXIII, vi, p. 703, 'On Herschel's Condition and the Optical Cosine Law.'

CHAPTER V

THE COMPUTATION OF OPTICAL SYSTEMS

1. *The Focal Eikonal for a Single Spherical Surface**. Hitherto the performance of the general symmetrical optical system has been considered and no specialisation has been made to particular systems; indeed it has not been assumed for the most part that the component surfaces are spherical; but only that they are surfaces of revolution. It has now to be seen how the preceding investigations may be linked up with special systems and to this end two steps are necessary: the aberration coefficients must be evaluated for a single surface, generally spherical; and then addition formulae must be obtained which will give the aberration coefficients of the whole system in terms of the coefficients of the component single surfaces.

Let U be the reduced eikonal when the principal foci are taken as base points; this is known as the 'reduced focal eikonal' and it is evident that it is a constant of the system. Moreover if $u - \Phi$ be the customary eikonal the following relation holds:

$$u - \Phi = U + \frac{1}{m}(1 - L) + m(1 - L'), \quad \ldots\ldots\ldots\ldots\ldots(1)$$

the notation being as in the preceding chapters. The evaluation of U, therefore, may be undertaken, instead of the evaluation of $u - \Phi$. The first application of this is to a single surface.

* Cf. *Proc. Camb. Phil. Soc.* XXIII, iv, 1926, p. 461.

2. Let A be the pole and O the centre of a spherical surface of radius r and modified power J, separating media of indices μ and μ', and let F_1 and F_2 be the principal foci of the surface: then

$$F_1A = \mu/\kappa \text{ and } AF_2 = \mu'/\kappa,$$

where
$$\kappa = (\mu' - \mu)/r.$$

Let PQP' be a ray whose direction cosines are L, M, N and L', M', N' respectively and let P and P' be the feet of the perpendiculars from F_1 and F_2; let Q be upon the refracting surface. Let U be the focal

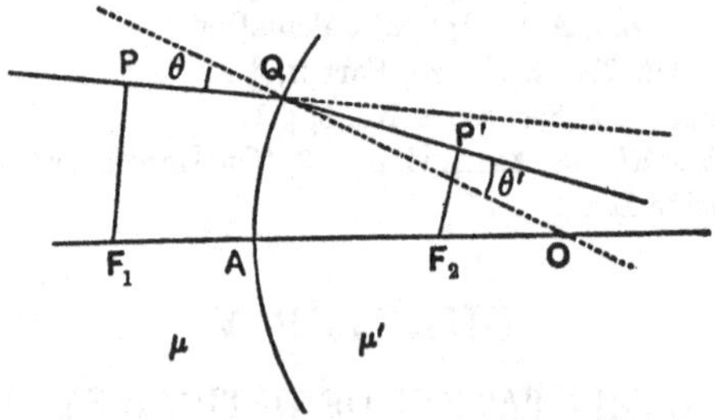

reduced eikonal; then if a constant be added to ensure the vanishing of U for an axial ray

$$U = J\{\mu(PQ - F_1A) + \mu'(QP' - AF_2)\}. \quad \ldots\ldots\ldots\ldots(1)$$

Now $\quad PQ = L \,.\, F_1O - r\cos\theta$ and $QP' = r\cos\theta' - L' \,.\, F_2O,$

where θ and θ' are the angles of incidence and refraction; so that

$$PQ = \frac{\mu' r L}{\mu' - \mu} - r\cos\theta \text{ and } QP' = r\cos\theta' + \frac{\mu r L'}{\mu' - \mu}.$$

Substituting in (1)

$$U = \frac{\mu' - \mu}{\mu\mu'}(\mu'\cos\theta' - \mu\cos\theta) + L + L' - \frac{(\mu' - \mu)^2}{\mu\mu'} - 2$$

$$= \frac{(\mu' - \mu)^2}{\mu\mu'}\left\{1 - \frac{2\mu\mu'}{(\mu' - \mu)^2}(LL' + MM' + NN' - 1)\right\}^{\frac{1}{2}}$$

$$+ L + L' - \frac{(\mu' - \mu)^2}{\mu\mu'} - 2. \quad \ldots\ldots(2)$$

Owing to the symmetry of the system this may be expressed in terms of the variables a, b and c of Chapter II, § 6, viz.,

$$a = M^2 + N^2, \quad b = MM' + NN' \text{ and } c = M'^2 + N'^2;$$

from which
$$L = \sqrt{1 - a} \text{ and } L' = \sqrt{1 - c}.$$

Then from (2) $$U = \frac{1}{v}(1+v\chi)^{\frac{1}{2}} + L + L' - \frac{1}{v} - 2, \qquad (3)$$

where $$v = \frac{\mu\mu'}{(\mu'-\mu)^2} \text{ and } \chi + 2(LL' + b - 1) = 0.$$

This expression (3) may be expanded in terms of a, b and c, and the coefficients of the various terms will involve simply the function v—an invariant of the system. For the consideration of the first order aberrations only the second order terms need be taken, i.e.,

$$U = -b - \tfrac{1}{8}\{v(a-2b+c)^2 + 2ca\}. \qquad (4)$$

3. *The Aberration Coefficients.* The focal eikonal has been evaluated now for a single spherical surface; that for a coaxial system of spherical surfaces may be written

$$U = -b - \tfrac{1}{8}\{A'a^2 - 4B'ab + 2(C+\varpi)ac + 4Cb^2 - 4Bbc + Ac^2\}, \qquad (1)$$

together with terms of higher orders. The coefficients A, B, C, ϖ, B' and A' are constants of the optical system, i.e., they are independent of the particular conjugate planes to be chosen and also of the pupil planes; and these are the quantities which it is convenient to evaluate. For, when they are known, simple relations will give the corresponding values of the aberration coefficients σ_1, σ_2, σ_3, σ_4, σ_5 and σ_6 in terms of them; because there will be a relation such as (1) § 1, i.e.,

$$u - \Phi = U + \frac{1}{m}(1-L) + m(1-L). \qquad (2)$$

The right-hand side of this equation is a known function of the variables a, b and c while the left-hand side is a function of α, β and γ; and we may eliminate the former variables by means of the relations given in Chapter II, § 6. Since (2) is an identity we may equate coefficients and we obtain the relations set out in the following scheme*:

	A	$-(B+\frac{1}{4})$	$(C+\varpi/3)$	$-(B'+\frac{1}{4})$	A'	$\frac{1}{4}(s-m)$
σ_1	1	$4m$	$6m^2$	$4m^3$	m^4	0
σ_2	1	$3m+s$	$3m(m+s)$	$m^2(m+3s)$	m^3s	$1-m^2$
$\frac{\sigma_3+2\sigma_4}{3}$	1	$2m+2s$	$m^2+4ms+s^2$	$2ms(m+s)$	m^2s^2	$2(1-ms)$
σ_5	1	$m+3s$	$3s(m+s)$	$s^2(3m+s)$	s^3m	$3(1-s^2)$
σ_6	1	$4s$	$6s^2$	$4s^3$	s^4	0

* Cf. T. Smith, *Trans. Opt. Soc.* XVIII, 1917.

For example,

$$\sigma_1 = A - 4m(B+\tfrac{1}{4}) + 6m^2(C+\varpi/3) - 4m^3(B'+\tfrac{1}{4}) + m^4 A'$$

is a typical result. Instead of eliminating a, b and c and so obtaining the σ's in terms of A, ... A', we may reverse the process and eliminate α, β and γ and obtain A, ... A' in terms of the σ's. In this way we find that, when all the first order aberrations vanish, for the conjugate planes at magnification m, the following relations must hold:

$$(A-m)/m^3 = B/m^2 = C/m = \varpi/0 = B' = A'm - 1 = -(s^3-m)/(s-m)^3.$$

Moreover it will be found that

$$\sigma_3 - \sigma_4 = \varpi(s-m)^2,$$

so that ϖ is the constant appearing in Chapter III, § 6. But a simpler method of evaluating ϖ is as follows: from the relations between the variables a, b and c and α, β and γ we have

$$4\frac{\partial^2}{\partial\gamma\partial\alpha} - \frac{\partial^2}{\partial\beta^2} = (s-m)^2\left(4\frac{\partial^2}{\partial c\partial a} - \frac{\partial^2}{\partial b^2}\right), \quad\text{..............(3)}$$

and applying this to relations (2) and (4) § 2

$$\sigma_3 - \sigma_4 = (s-m)^2,$$

where now the aberration coefficients refer only to a single surface. The curvature increment (P) (modified and reduced) produced by this single surface is therefore unity, i.e.,

$$\Delta\left(\frac{1}{\mathrm{P}}\right) = 1 \quad\text{or}\quad \Delta\left(\frac{1}{\mu\rho}\right) = J = \frac{\kappa}{\mu\mu'},$$

where now ρ is the geometrical radius of curvature of the object (or image) surface; and the curvature increment produced by the system of surfaces is

$$\Delta\left(\frac{1}{\mu\rho}\right) = \Sigma J = \Sigma\frac{\kappa}{\mu\mu'} = (J\varpi)_{1,n}. \quad\text{..................(4)}$$

ϖ is the Petzval sum and the condition for flatness of field is

$$\Sigma\kappa/\mu\mu' = 0. \quad\text{............................(5)}$$

4. *The Equation for the Addition of Aberrations.* A general relation is needed between the aberration coefficients for a combined system and those for the several components and it is obtained as follows. We consider n coaxial systems, the reduced eikonal for system λ, of modified power J_λ, being written $(u-\Phi)_\lambda$, the corresponding functions for the combined system being written $J_{1,n}$ and $(u-\Phi)_{1,n}$. The base points used are successively conjugate axial points for the several systems,

the image point for system λ being the object point for system $\lambda+1$. Then using the figure and notation of §§ 5 and 6, Chapter II, we have, for the component λ,

$$\left(\frac{u-\Phi}{J}\right)_\lambda = E_\lambda - e_\lambda = \int_{Q_{\lambda-1}}^{Q'_\lambda} \mu\, ds - \int_{O_{\lambda-1}}^{O'_\lambda} \mu\, ds,$$

$Q_{\lambda-1}$ and Q'_λ being the feet of the perpendiculars upon the ray from the conjugate base points $O_{\lambda-1}$ and O_λ. Thus

$$\sum_{\lambda=1}^{n}\left(\frac{u-\Phi}{J}\right)_\lambda = \sum_{\lambda=1}^{n}\int_{Q_{\lambda-1}}^{Q'_\lambda} \mu\, ds - \sum_{\lambda=1}^{n}\int_{O_{\lambda-1}}^{O'_\lambda} \mu\, ds$$

$$= \int_{Q_0}^{Q'_n} \mu\, ds - \int_{O_0}^{O'_n} \mu\, ds = \left(\frac{u-\Phi}{J}\right)_{1,n},$$

since Q'_λ and Q_λ and also O'_λ and O_λ coincide for all values of λ. Thus we have the general equation for the addition of aberrations in the form

$$\left(\frac{u-\Phi}{J}\right)_{1,n} = \sum_{\lambda=1}^{n}\left(\frac{u-\Phi}{J}\right)_\lambda. \quad \text{.....................(1)}$$

A result, of use subsequently, may be obtained here; let two coaxial systems, of modified powers J_1 and J_2 and principal foci F_1, F_2 and F_1', F_2', combine to form a system of modified power J and principal foci F, F' (§ 12, Chapter I); and let P, P' and P'' be three corresponding axial points, i.e., P' is the image of P in the first system and P'' that of P' in the second sysetm, and let the corresponding magnifications be m_1 and m_2 so that m $(= m_1 m_2)$ is the magnification for the combined system. Then

$$F'P'' = F'F_2' + F_2'P''$$

and, using the relations of § 12, Chapter I, we have

$$m_2 J - J_1 - m J_2 = 0. \quad \text{......................(2)}$$

5. *Addition of Aberration Coefficients.* A direct application of this has now to be made to the various aberration coefficients; two systems may be considered and dashes used to distinguish them. Three sets of variables $a\ldots$, $a'\ldots$, $a''\ldots$ may be defined as in Chapter II, § 6, i.e.,

$$a(s-m)^2 = (M - sM'')^2 + (N - sN'')^2, \text{ etc.,}$$

$$a'(s'-m')^2 = (M - s'M')^2 + (N - s'N')^2, \text{ etc.,}$$

$$a''(s''-m'')^2 = (M' - s''M'')^2 + (N' - s''N'')^2, \text{ etc.,}$$

where capital letters are used to denote direction cosines, and

α' and α'' refer to the single systems and α to the compound system Also we have

$$m = m'm'' \text{ and } s = s's''. \quad \text{......................(1)}$$

Now from the preceding paragraph the fundamental equation is

$$\frac{u-\Phi}{J} - \frac{u'-\Phi'}{J'} - \frac{u''-\Phi''}{J''} = 0; \quad \text{..................(2)}$$

so that

$$\frac{\partial}{\partial M}\left\{\frac{u-\Phi}{J} - \frac{u'-\Phi'}{J'}\right\} = 0,$$

since the third term of (2) does not involve the variable M explicitly: and to the first order this is

$$\frac{\partial}{\partial M}(u/J - u'/J') = 0,$$

i.e.,

$$\frac{1}{J}\frac{\partial u}{\partial \gamma}\frac{\partial \gamma}{\partial M} - \frac{1}{J'}\frac{\partial u'}{\partial \gamma'}\frac{\partial \gamma'}{\partial M} = 0.$$

Remembering now the first approximation to u, obtained in Chapter II, § 9, viz.,

$$2mu = (s-m)^2\,\gamma,$$

this result is

$$\frac{1}{J}\frac{(s-m)^2}{2m} \cdot \frac{2(M - mM'')}{(s-m)^2} - \frac{1}{J'}\frac{(s'-m')^2}{2m'} \cdot \frac{2(M - m'M')}{(s'-m')^2} = 0,$$

i.e.,

$$\frac{M - mM''}{mJ} = \frac{M - m'M'}{m'J'}. \quad \text{......................(3)}$$

Similarly we may differentiate (2) with respect to M' and obtain

$$\frac{\partial}{\partial M'}\left(\frac{u'}{J'} + \frac{u''}{J''}\right) = 0,$$

to the first order; from which follows the relation

$$\frac{M - m'M'}{J'} = \frac{M' - m''M''}{m''J''}. \quad \text{..................(4)}$$

Combining (3) and (4) we have

$$\frac{M - m'M'}{J'} = \frac{M' - m''M''}{m''J''} = \frac{M - mM''}{m''J}, \quad \text{............(5)}$$

and a similar relation involving N, N' and N'': we have also two corresponding relations involving s, s' and s'' in place of m, m' and m''. Again, corresponding to (2) § 4 will be a relation involving s instead of m, viz.

$$s''J - J' - sJ'' = 0,$$

and this, combined with (2) § 4, may be solved for the modified powers, giving

$$\frac{J}{s-m}=\frac{J'}{sm''-s''m}=\frac{J''}{s''-m''}=\frac{J'}{m''s''(s'-m')}. \quad \text{.........(6)}$$

Thus
$$\frac{a(s-m)^2}{a''(s''-m'')^2}=\frac{(M-sM'')^2+(N-sN'')^2}{(M'-s''M'')^2+(N'-s''N'')^2}=\left(\frac{J}{J''}\right)^2,$$

i.e., $a=a''$ from (5) and (6). Similarly

$$\frac{a(s-m)^2}{a'(s'-m')^2}=\frac{(M-sM'')^2+(N-sN'')^2}{(M-s'M')^2+(N-s'N')^2}=\left(\frac{s''J}{J'}\right)^2,$$

so that $a'=m''^2a$. In the same way we have the relations

$$\beta=\beta'',\ \gamma=\gamma'',\ \beta'=s''m''\beta \text{ and } \gamma'=s''^2\gamma. \quad \text{.........(7)}$$

6. Let us consider now the equation for the addition of aberrations (1) § 4; it is

$$(u-\Phi)/J-(u-\Phi)'/J'-(u-\Phi)''/J''=0; \quad \text{.........(1)}$$

the u's are functions only of the γ's, which will vary slightly with the introduction of aberrations; the first approximations are given in § 5.

If we write

$$\Psi\equiv u/J-u'/J'-u''/J'', \quad \text{.....................(2)}$$

then, by using (6) and (7) § 5, combined with the expansion for u, it is seen that the first and second order terms in Ψ vanish identically. Moreover the variation of Ψ brought about by the introduction of aberrations is given by

$$\delta\Psi \propto m''(\delta M-m\delta M'')-m''(\delta M-m'\delta M')-m(\delta M'-m''\delta M'')$$
$$+m''(\delta N-m\delta N'')-m''(\delta N-m'\delta N')-m(\delta N'-m''\delta N'')$$

i.e.,
$$\delta\Psi=0,$$

as far as terms involving γ^2. So far, therefore, as second order aberrations are concerned (1) reduces to

$$\Phi/J-\Phi'/J'-\Phi''/J''=0, \quad \text{.......................(3)}$$

which accordingly may be used to give the addition formulae for the primary aberrations.

In conformity with the notation of Chapter III, § 3, we may write

$$8\Phi=\sigma_1a^2-4\sigma_2a\beta+2\sigma_3a\gamma+4\sigma_4\beta^2-4\sigma_5\beta\gamma+\sigma_6\gamma^2;$$

substituting therefore in (3), and using (7) § 5, we may equate coefficients, since (3) is an identity, and obtain

$$\left(\frac{\sigma_1}{J}\right)=\left(\frac{\sigma_1}{J}\right)'m''^4+\left(\frac{\sigma_1}{J}\right)'',\quad \left(\frac{\sigma_2}{J}\right)=\left(\frac{\sigma_2}{J}\right)'m''^3s''+\left(\frac{\sigma_2}{J}\right)'',$$

$$\left(\frac{\sigma_3}{J}\right)=\left(\frac{\sigma_3}{J}\right)'m''^2s''^2+\left(\frac{\sigma_3}{J}\right)'',\quad \left(\frac{\sigma_4}{J}\right)=\left(\frac{\sigma_4}{J}\right)'m''^2s''^2+\left(\frac{\sigma_4}{J}\right)'',$$

$$\left(\frac{\sigma_5}{J}\right)=\left(\frac{\sigma_5}{J}\right)'m''s''^3+\left(\frac{\sigma_5}{J}\right)'',\quad \left(\frac{\sigma_6}{J}\right)=\left(\frac{\sigma_6}{J}\right)'s''^4+\left(\frac{\sigma_6}{J}\right)''.$$

These results may clearly be extended to a composite system formed by combining n systems—which as a special case may themselves be simple surfaces; then if $m_{p,q}$ denote the magnification for the combined systems from p to q inclusive, and the like for $s_{p,q}$, the general formulae for the addition of aberrations follow, viz.,

$$\left(\frac{\sigma_1}{J}\right)_{1,n} = \sum_{\lambda=1}^{n} \left(\frac{\sigma_1}{J}\right)_\lambda m^4_{\lambda+1,n}, \quad \left(\frac{\sigma_2}{J}\right)_{1,n} = \sum_{\lambda=1}^{n} \left(\frac{\sigma_2}{J}\right)_\lambda m^3_{\lambda+1,n} s_{\lambda+1,n},$$

$$\left(\frac{\sigma_3}{J}\right)_{1,n} = \sum_{\lambda=1}^{n} \left(\frac{\sigma_3}{J}\right)_\lambda m^2_{\lambda+1,n} s^2_{\lambda+1,n}, \quad \left(\frac{\sigma_4}{J}\right)_{1,n} = \sum_{\lambda=1}^{n} \left(\frac{\sigma_4}{J}\right)_\lambda m^2_{\lambda+1,n} s^2_{\lambda+1,n},$$

$$\left(\frac{\sigma_5}{J}\right)_{1,n} = \sum_{\lambda=1}^{n} \left(\frac{\sigma_5}{J}\right)_\lambda m_{\lambda+1,n} s^3_{\lambda+1,n}, \quad \left(\frac{\sigma_6}{J}\right)_{1,n} = \sum_{\lambda=1}^{n} \left(\frac{\sigma_6}{J}\right)_\lambda s^4_{\lambda+1,n} {}^{*}.$$

From this result, therefore, the optical properties of a complete system may be written down at once from a knowledge of the corresponding properties of the several components of the system, as far as first order aberrations are concerned.

7. *Formulae for Aberration Coefficients.* Let us now consider a single spherical surface for which the focal eikonal is given in § 2; as in § 1 we have the relation

$$u - \Phi = U + (1 - L)/m + (1 - L')\,m, \quad \ldots\ldots\ldots\ldots\ldots(1)$$

in which the left-hand side is a function of the variables α, β and γ and the right-hand side of the variables a, b and c; and linear relations exist between these variables, given in Chapter II, § 6. (1) is an identity and we may therefore substitute either set of variables in terms of the other set and equate the coefficients of the corresponding powers. Let us eliminate a, b and c; then remembering the form of Φ and making use of the expansion of u in terms of γ we have

$$\sigma_1 = (m-1)^2 \{v\,(m-1)^2 - m\},$$

$$\sigma_2 = (m-1)(s-1)\{v\,(m-1)^2 - m\},$$

$$\sigma_3 = v\,(m-1)^2 (s-1)^2 + (m-1)(m-s^2)$$

$$\sigma_4 = (s-1)^2 \{v\,(m-1)^2 - m\},$$

$$\sigma_5 = v\,(m-1)(s-1)^3 + (s-1)(m-s^2),$$

and $$\sigma_6 = (s-1)^2\{v\,(s-1)^2 - s\}; \quad \ldots\ldots\ldots\ldots\ldots\ldots(2)$$

where the suffix λ may be attached to each quantity since these relations refer to the surface λ. Moreover

$$m_\lambda m_{\lambda+1,n} = m_{\lambda,n} \text{ and } m_{\lambda+1,n} J_{1,n} = J_{1,\lambda} + m J_{\lambda+1,n}; \quad \ldots(3)$$

* Cf. also *Proc. Camb. Phil. Soc.* XXIII, v, p. 584.

and similar relations hold involving s. Here m denotes the magnification for the combined system. On substitution in the first of (2) we have

$$J^4_{1,n}(\sigma_1)_\lambda m^4_{\lambda+1,n} = \{J_{1,\lambda-1} - J_{1,\lambda} + m(J_{\lambda,n} - J_{\lambda+1,n})\}^2 \times$$
$$[v_\lambda\{J_{1,\lambda-1} - J_{1,\lambda} + m(J_{\lambda,n} - J_{\lambda+1,n})\}^2 - (J_{1,\lambda-1} + mJ_{\lambda,n})(J_{1,\lambda} + mJ_{\lambda+1,n})]. \quad \ldots(4)$$

Now if we write $m = 0$ in the table of § 3, it appears that $\sigma_1 = A$; so that, on making use of the addition formulae of § 6, we have

$$J^3_{1,n}A = \sum_{\lambda=1}^{n} \frac{1}{J_\lambda}\left(J_\lambda \frac{\partial J_{1,\lambda}}{\partial J_\lambda}\right)^2 \left\{v_\lambda \left(J_\lambda \frac{\partial J_{1,\lambda}}{\partial J_\lambda}\right)^2 - J_{1,\lambda-1}J_{1,\lambda}\right\} \ldots\ldots(5)$$

Again on writing $m = 0$ in the table of § 3 it is seen that $\sigma_2 + sB = A$; so that we may evaluate the coefficient B, and similarly the remaining coefficients, obtaining results corresponding to (5). The most useful case is that in which the end media are air, of refractive index unity; and then expressing the formula (5) in terms of the ordinary powers we have

$$K^3_{1,n}A = \sum_{\lambda=1}^{n} \frac{\partial K_{1,\lambda}}{\partial \kappa_\lambda}\left(\frac{K_{1,\lambda}}{\mu^2_\lambda} - \frac{K_{1,\lambda-1}}{\mu^2_{\lambda-1}}\right)\left(\mu_\lambda R_\lambda \frac{\partial K_{1,\lambda}}{\partial \kappa_\lambda} - K_{1,\lambda}\right)^2 \quad (6)$$

using the results given in Chapter I, § 9; here R_λ is the curvature of surface λ. The results may be expressed concisely as follows; let

$$K^3_{1,n}k_p = \frac{K_{1,p}}{\mu^2_p} - \frac{K_{1,p-1}}{\mu^2_{p-1}}, \quad K^3_{1,n}k'_p = \frac{K_{p,n}}{\mu^2_{p-1}} - \frac{K_{p+1,n}}{\mu^2_p},$$

$$x_p = \mu_p R_p \frac{\partial K_{1,p}}{\partial \kappa_p} - K_{1,p}, \quad x'_p = \mu_p R_p \frac{\partial K_{p,n}}{\partial \kappa_p} + K_{p+1,n};$$

then

$$A = \sum_{\lambda=1}^{n} k_\lambda x^2_\lambda \frac{\partial K_{1,\lambda}}{\partial \kappa_\lambda}, \quad B = \sum_{\lambda=1}^{n} k_\lambda x_\lambda x'_\lambda \frac{\partial K_{1,\lambda}}{\partial \kappa_\lambda}, \quad C = \sum_{\lambda=1}^{n} k_\lambda x'^2_\lambda \frac{\partial K_{1,\lambda}}{\partial \kappa_\lambda},$$

$$C' = \sum_{\lambda=1}^{n} k'_\lambda x^2_\lambda \frac{\partial K_{\lambda,n}}{\partial \kappa_\lambda}, \quad B' = \sum_{\lambda=1}^{n} k'_\lambda x_\lambda x'_\lambda \frac{\partial K_{\lambda,n}}{\partial \kappa_\lambda}, \quad A' = \sum_{\lambda=1}^{n} k'_\lambda x'^2_\lambda \frac{\partial K_{\lambda,n}}{\partial \kappa_\lambda},$$

$$K_{1,n}\varpi_{1,n} = \sum_{\lambda=1}^{n} \kappa_\lambda/\mu_{\lambda-1}\mu_\lambda. \quad \ldots\ldots\ldots\ldots\ldots\ldots\ldots(7)$$

These summations may be calculated readily for they involve quantities similar to those used in Chapter I to obtain the power of the combined system; and when $A, B, \ldots A'$ are known—being constants for a given system—the aberration coefficients $\sigma_1, \ldots \sigma_6$ may be obtained from the table of § 3. These will then give the first order aberrations for any conjugate planes and any exit-pupil.

As a check upon the result (4) § 3 we have from (2)* above

$$\varpi = \frac{\sigma_3 - \sigma_4}{(s-m)^2} = \frac{J_{1,n}}{(s-m)^2} \sum_{\lambda=1}^{n} \left(\frac{\sigma_3 - \sigma_4}{J}\right)_\lambda m^2_{\lambda+1,n} s^2_{\lambda+1,n}$$

$$= \frac{J_{1,n}}{(s-m)^2} \sum_{\lambda=1}^{n} \frac{(s-m)_\lambda^2}{J_\lambda} m^2_{\lambda+1,n} s^2_{\lambda+1,n}$$

$$= J_{1,n} \sum_{\lambda=1}^{n} (J_{1,\lambda} J_{\lambda,n} - J_{\lambda+1,n} J_{\lambda,n})^2 / J_\lambda J^4_{1,n}$$

$$= J_{1,n} \sum_{\lambda=1}^{n} J_\lambda / J^2_{1,n} = \frac{1}{J_{1,n}} \sum_{\lambda=1}^{n} J_\lambda,$$

i.e.,

$$(J\varpi)_{1,n} = \sum_{\lambda=1}^{n} J_\lambda.$$

Or, if the end media be air,

$$(K\varpi)_{1,n} = \sum_{\lambda=1}^{n} \kappa_\lambda / \mu_{\lambda-1} \mu_\lambda .$$

8. *Aberrations at the Principal Foci.* The formulae of the preceding paragraph give completely the primary aberrations of an optical system—or rather they give the constants $A, \ldots A'$ in terms of which the aberration coefficients are known. The quantities involved in the formulae are the same as occur in the formulae of Chapter I, so that very little additional work is necessary for their calculation; and in fact with a machine their computation is quite rapid. The aberrations of higher orders may be dealt with similarly but they are beyond the scope of this tract. For these reference may be made to a paper by T. Smith†. Similar formulae may be found for them involving again the same quantities and their computation may be carried out in the same way.

It will be observed that if we write $m=0$ we are dealing with the second principal focus and we have from § 3

$$\sigma_1 = A, \; \sigma_2 = A - sB, \quad \ldots\ldots\ldots\ldots\ldots\ldots\ldots(1)$$

i.e.,

$$\sigma_2 = A - B,$$

if $s=1$ and the exit-pupil is at the second unit plane; this very commonly is the case. Thus $A=0$ implies the absence of spherical aberration at the second principal focus while if, in addition, $B=0$ first order coma also has been eliminated, i.e., the condition $B=0$ implies the satisfaction of the sine-condition for the second principal focus, to this order of approximation.

In the case of thin systems $s=1$ implies that the pupil-planes coincide with the boundary of the optical surfaces; but here special relations hold between the six constants $A, \ldots A'$.

* By the application of the results (2) of this paragraph it may be verified that the various forms of conditions commonly given are at once deducible from the formulae of Chapter III: e.g. Whittaker, *Camb. Math. Tract*, No. 7.

† 'The Addition of Aberrations,' T. Smith, *Trans. Opt. Soc.* XXV, No. 4.

9. *Thin Systems.* In practice 'thin' systems will not occur but cases arise in which the properties of a system may be investigated with some accuracy by regarding it as thin. And here the constants $A, \dots A'$ will not be independent; we propose to find the relations between them. Without loss of generality we may write $K_{1,n} = 1$ and then

$$A = \sum_1^n k_p x_p^2, \quad B = \sum_1^n k_p x_p x_p', \quad C = \sum_1^n k_p x_p'^2 = \sum_1^n k_p' x_p^2;$$

and $$A' = \sum_1^n k_p' x_p'^2, \quad B' = \sum_1^n k_p' x_p x_p', \quad \varpi = \sum_1^n \kappa_p / \mu_{p-1}\mu_p;$$

where $$k_p = \left(\frac{K_{1,p}}{\mu_p^2} - \frac{K_{1,p-1}}{\mu_{p-1}^2}\right), \quad k_p' = \left(\frac{K_{p,n}}{\mu_{p-1}^2} - \frac{K_{p+1,n}}{\mu_p^2}\right),$$

and $$x_p = \mu_p R_p - K_{1,p}, \quad x_p' = \mu_p R_p + K_{p+1\ n}.$$

Thus

$$\begin{aligned} A - 2B + C &= \sum_1^n k_p (x_p^2 - 2x_p x_p' + x_p'^2) = \sum_1^n k_p (x_p - x_p')^2 \\ &= \sum_1^n k_p (K_{1,p} + K_{p+1,n})^2 = \sum_1^n k_p \\ &= \left(\frac{\kappa_1}{\mu_1^2} + \frac{\kappa_1 + \kappa_2}{\mu_2^2} - \frac{\kappa_1}{\mu_1^2} + \dots + \frac{\kappa_1 + \dots + \kappa_n}{\mu_n^2} - \frac{\kappa_1 + \dots + \kappa_{n-1}}{\mu_{n-1}^2}\right) \\ &= \frac{1}{\mu_n^2}. \qquad \text{.........(1)} \end{aligned}$$

Similarly $$A' - 2B' + C = \frac{1}{\mu_0^2}. \qquad \text{.........(2)}$$

And

$$\begin{aligned} B + B' - 2C &= \sum_1^n (k_p' x_p - k_p x_p') = \sum_1^n \left\{x_p \left(\frac{1}{\mu_{p-1}^2} - \frac{1}{\mu_p^2}\right) - k_p\right\} \\ &= \sum_1^n \left\{\mu_p R_p \left(\frac{1}{\mu_{p-1}^2} - \frac{1}{\mu_p^2}\right) - \frac{\kappa_p}{\mu_{p-1}^2}\right\} = \sum_1^n \frac{\kappa_p}{\mu_{p-1}\mu_p} \\ &= \varpi. \qquad \text{.........(3)} \end{aligned}$$

Thus if the end media be air we have the relations

$$\left.\begin{aligned} A + C &= 2B + 1 \\ B + B' &= 2C + \varpi \\ A' + C &= 2B' + 1 \end{aligned}\right\}; \qquad \text{.........(4)}$$

so that for a thin system three only of the constants need be evaluated.

REFERENCES

Smith, T. *Dictionary of Applied Physics*, Art. 'Optical Calculations.'

—— *Trans. Opt. Soc.* XVIII, 1917.

—— *Collected Researches of the National Phys. Lab.* vols. XVI, XVII.

Steward, G. C. *Phil. Trans. Roy. Soc.* A, 225, 1925, Part I.

—— *Proc. Camb. Phil. Soc.* XXIII, Part IV, p. 461.

—— *Proc. Camb. Phil. Soc.* XXIII, Part V, p. 584.

—— 'On Invariant and Semi-Invariant Aberrations of the Symmetrical Optical System.' *Acta Math.* vol. 67 (1936).

CHAPTER VI

DIFFRACTION PATTERNS ASSOCIATED WITH THE SYMMETRICAL OPTICAL SYSTEM

1. *General Formulae.* The preceding discussion of the working of the symmetrical optical system has been purely geometrical in character and it has been assumed that the only requisite for a point image is that the emergent rays should themselves pass through a point. But owing to the physical nature of light and to its finite, though small, wave-length this assumption is untenable; for a part only of the complete incident wave will pass through the system, and for this reason diffraction phenomena will appear; so that corresponding to a point source of light there will in no case be a point image but, instead, a diffraction pattern. And this diffraction pattern will be modified profoundly by the presence of the geometrical aberrations which have been considered in the preceding chapters. In the absence of these aberrations, i.e., if we assume perfect definition for two conjugate planes, the ideal diffraction pattern is a series of concentric circles, alternately bright and dark, with centre at the conjugate point; this was investigated by Airy in 1834. We desire to examine this pattern and, in particular, the modification of it produced by the geometrical aberrations of the system.

The general method of procedure, as is usual in discussions of diffraction phenomena, is to apply Huyghens' Principle; which enables us to determine the effect, at a given point, of any propagated disturbance, when a surface surrounding the point can be taken, the secondary surface, over which the disturbance is known. It is assumed that each element of this secondary surface originates a disturbance, the phase and amplitude of which are appropriate to the element considered.

2. The emergent wave is limited as to size and shape by the exit-pupil of the system and we may take this as the secondary surface, the disturbance over which produces the diffraction effects; and by integrating over this surface the diffraction pattern may be obtained.

Let AA' be the axis of a symmetrical optical system and let O, O'

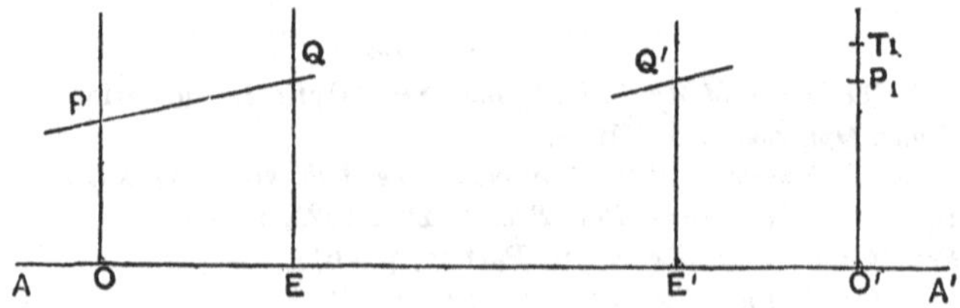

and E, E' be pairs of conjugate axial points. Let a ray cut the normal planes through O, E and E' in P, Q and Q' respectively, and let P_1 upon the normal plane O' be the Gaussian conjugate of P; the planes E and E' will be taken as the pupil planes of the system. In conformity with the preceding notation the co-ordinates of P and Q' may be taken to be Y, Z and y', z', and those of P_1, Y_1, Z_1. Let

$$V = \int_P^{Q'} \mu\, ds,$$

so that V is the characteristic function, and $2\pi V/\lambda$ is the phase difference between the luminous vibrations at the points P and Q'; where λ is the wave-length of the disturbance measured *in vacuo* or, of course, the reduced length measured in any medium. The phase contribution of the element dS of the exit-pupil at Q' to the disturbance at a point T_1 upon the normal plane through O' and near to P_1 is

$$2\pi\left(V + r + \frac{\lambda}{4}\right)\Big/\lambda^*,$$

where $Q'T_1 = r$; so that the intensity of the disturbance at T_1 is proportional to the squared modulus of the expression

$$\int e^{\frac{2\pi i}{\lambda}(V+r)}\, dS\dagger. \qquad \text{.........................(1)}$$

The amplitude of the vibration has been omitted and this is justified by the consideration that T_1 is near to P_1 and the emergent wave surfaces are approximately concentric spheres with centre at P_1, so that the amplitude enters as a factor of the form A/r, where A is a constant and r, to the first order, is equal to the distance between the planes E' and O'.

3. Now the function V depends only upon the variables θ, ϕ and ψ and the variables of integration over the pupil planes are y' and z': so that θ is a constant as far as this integration is concerned. Moreover we have

$$V = f(\theta) - d\sqrt{1+u} + F, \qquad \text{....................(1)}$$

where F is the aberration function. Let ρ', ϕ' be the polar co-ordinates of T_1 referred to parallel axes through P_1 and let ρ, ϕ_1 be the polar co-ordinates of Q' upon the exit-pupil: then if $E'O' = d$

$$r^2 = d^2 + (Y_1 + \rho' \cos\phi' - \rho\cos\phi_1)^2 + (Z_1 + \rho'\sin\phi' - \rho\sin\phi_1)^2,$$

i.e.,

$$r = d\sqrt{1 + u + \alpha},$$

if we write $Z_1 = 0$, where

$$\alpha d^2 = 2\rho' Y_1 \cos\phi' - 2\rho\rho'\cos(\phi_1 - \phi') + \rho'^2 \quad \text{and} \quad u = \theta - \phi + \psi:$$

* Rayleigh, *Sci. Papers*, vol. III, p. 74.

† We must take λ and r to be modified distances, i.e., distances multiplied by the modified power of the optical system: cf. p. 27, footnote.

and ρ' is small so that to a first approximation we have from (1)

$$V + r = f(\theta) + \frac{\rho' Y_1}{d} \cos\phi' - \frac{\rho\rho'}{d} \cos(\phi_1 - \phi') + F. \quad \ldots\ldots(2)$$

The light intensity is required not only at neighbouring points upon the normal plane $O'P_1$ but also at near points upon neighbouring planes; for we have to consider the effect of change of focus. We may consider a normal plane distant X from O', measured positively in the direction $O'E'$, where X will be small; this is equivalent to writing $(d - X)$ in place of d above and the additional terms obtained are therefore

$$-X + \frac{X}{2d^2}(Y_1^2 - 2\rho Y_1 \cos\phi_1 + \rho^2), \quad \ldots\ldots\ldots\ldots\ldots(3)$$

to the first approximation. Thus, finally, we have

$$V + r = \text{constant} + \Psi,$$

where $$\Psi = F - (\rho\rho'/d)\cos(\phi_1 - \phi') + \frac{X}{2d^2}(\rho^2 - 2\rho Y_1 \cos\phi_1), \quad \ldots\ldots(4)$$

and the terms not involving the variables of integration ρ, ϕ_1 are absorbed in the constant. Remembering that a constant real term in (4) does not affect the modulus of (1) § 2, that result becomes

$$\int e^{i\kappa\Psi}\, dS, \quad \ldots\ldots\ldots\ldots\ldots\ldots\ldots\ldots\ldots(5)$$

where $\kappa\lambda = 2\pi$. The squared modulus of this is proportional to the light intensity; if therefore we consider the value of (5) at the Gaussian image P_1 and in the absence of aberrations, i.e., $F = 0$, $\rho' = 0$ and $X = 0$ and therefore $\Psi = 0$, we have simply

$$\int dS = S,$$

where S is the area of the exit-pupil. The result

$$I = \left| \frac{1}{S} \int e^{i\kappa\Psi}\, dS \right|^2 \quad \ldots\ldots\ldots\ldots\ldots\ldots\ldots(6)$$

gives therefore the light intensity at *any* point T_1, in the neighbourhood of P_1, relative to the intensity at P_1 for an aberrationless system, i.e., this intensity is taken as unity. The expression (6) is therefore the general result covering the case of exit-pupils of any form; in the present chapter we shall assume that the exit-pupil is circular with centre upon the axis of the system.

4. The co-ordinates of T_1 relative to P_1 as origin are ρ', ϕ'; it is convenient to use a slight modification of these and to introduce a new co-ordinate, the meaning of which is considered here. Let S be the exit-pupil with centre E' and radius ρ_1, and let P be a point upon a plane normal to and intersecting the axis of the system in O', where $E'O' = d$; let AB be the diameter of S parallel to $O'P$ and let $O'P = Y_1$. Then if Y_1 and ρ_1 be small compared with d we have approximately $AP \sim BP = 2Y_1\rho_1/d$, so that the difference in phase at P between rays

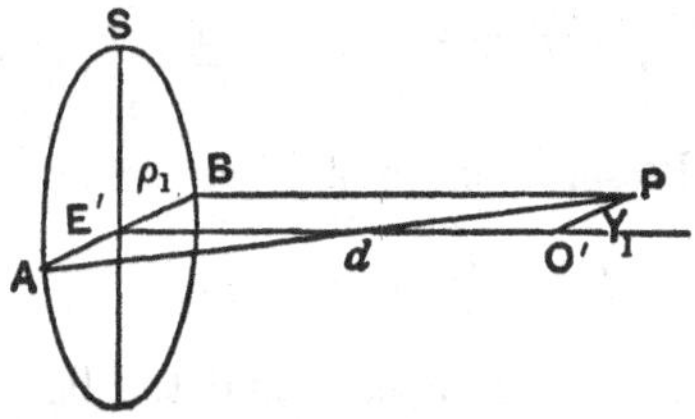

from A and B is $2\kappa Y_1\rho_1/d$, or $2z$, if we write $zd = \kappa Y_1\rho_1$; so that z may be used as a co-ordinate upon the image plane.

Again, $BO' = \sqrt{d^2 + \rho_1^2} = d\sqrt{1+\psi}$, so that in the case of an axial object point, i.e., $Y_1 = 0$ in (2) § 3, implying that $\theta = \phi = 0$, the optical path to O' is $V + BO'$; i.e., it is $f(0) + F$. The aberration function F therefore gives the excess phase at O' of the ray from B over that for the axial ray; and the only terms remaining in F are those depending upon the function ψ. We may write

$$\kappa F = 2\pi n_s, \quad \text{.............................(1)}$$

where n_s is the number of wave-lengths excess produced by aberration of order s, depending therefore upon a term containing ψ^{s+1} in the function F. Moreover if the emergent ray cut the axis at a distance X from O' we have

$$-\rho_1 X/d = D_{Y_1} = 2\rho_1 \frac{\partial F}{\partial \psi}\Big/ d,$$

where D_{Y_1} is the distance from O' of the point of intersection of the emergent ray with the normal plane $O'P$,

i.e., $$-\psi X/2 = \psi \frac{\partial F}{\partial \psi}.$$

So that if we write $2\pi n = \dfrac{X}{2}\left(\dfrac{\rho_1}{d}\right)^2 \dfrac{2\pi}{\lambda}$ we have

$$n + (s+1)\, n_s = 0, \quad \text{..........................(2)}$$

and this result gives the position of the marginal focus in the presence of axial aberration of order s.

5. *The Airy Disc.* We proceed to make specialisations of the general equation (6) § 3, and in the first place we discuss the 'ideal' diffraction pattern, i.e., in (4) § 3 we write $F=0$ and $X=0$ and so consider the light effect upon the conjugate plane in the absence of geometrical aberrations. Here

$$\kappa\Psi = -\frac{\kappa\rho\rho'}{d}\cos(\phi_1-\phi'),$$

so that we must consider the expression

$$\int_0^{\rho_1}\rho\,d\rho\int_0^{2\pi}e^{-i\frac{\kappa\rho\rho'}{d}\cos(\phi_1-\phi')}\,d\phi_1=\frac{1}{2\pi}\int_0^{\rho_1}\rho\,d\rho\,J_0\left(\frac{\kappa\rho\rho'}{d}\right),$$

where $J_0(t)$ is the Bessel function of order zero;

so that $$\sqrt{I}=\frac{2}{z^2}\int_0^z tJ_0(t)\,dt=2J_1(z)/z. \qquad (1)$$

The intensity depends only upon z and the light contours (lines of equal intensity) are therefore circles with the point P_1 as centre; the central intensity is of course taken as unity and we have therefore dark rings of zero intensity at the values of z given by the roots of the equation

$$J_1(z)=0, \qquad (2)$$

other than $z=0$; while the maxima are given by the roots of the equation in z

$$\frac{d}{dz}\left(\frac{J_1(z)}{z}\right)=0,$$

i.e., $$J_2(z)=0. \qquad (3)$$

If $z=z_1$ be a root of (2) or (3) the radius ρ' of the corresponding dark or bright ring is given by

$$2\pi\rho_1\rho'=z_1\lambda d,$$

so that the radii of the rings vary directly as the wave-length of the luminous vibration considered and inversely as the radius of the exit-pupil: the diffraction pattern therefore will commonly be coloured and we shall obtain zero intensities only if monochromatic light be used—or if the pattern be viewed through a colour filter.

It is of interest to investigate the proportion of the total intensity in the various rings, and to this end we must evaluate the expression

$$\int_0^z Iz\,dz=4\int_0^z J_1^2(z)\,.\,z^{-1}\,dz$$
$$=2\{1-J_0^2(z)-J_1^2(z)\}.$$

In general, therefore, the fraction of the illumination to be found *outside* any ring of radius z is

$$J_0^2(z)+J_1^2(z)$$

so that the fraction outside any dark ring is simply $J_0^2(z)$. From this it appears that more than 90°/$_\circ$ of the illumination is concentrated within the area of the second dark ring.

For large values of z we have

$$2\frac{J_1(z)}{z}=\frac{2}{z}\sqrt{\frac{2}{\pi z}}\sin\left(z-\tfrac{1}{4}\pi\right),$$

so that the maxima and minima occur at equal intervals; moreover the mean brightness varies as $1/z^3$*.

6. *Change of Focus and Spherical Aberration.* We wish to examine the effect of axial aberrations and we write $Y_1=0$ in (4) § 3; so that

$$\kappa\Psi=-\frac{\kappa\rho\rho'}{d}\cos(\phi_1-\phi')+\frac{\kappa X\rho^2}{2d^2}+\kappa F,$$

and F now will contain only terms in the variable ψ or $(\rho/d)^2$. If now

$$\kappa\rho\rho'=td,\quad \kappa X\rho^2=2\mu t^2d^2,\quad \kappa A_{0,0,s}\rho^{2s}=\nu_s t^{2s}d^{2s},$$

we have $$\kappa\Psi=-t\cos(\phi_1-\phi')+\mu t^2+\sum_{s=2}\nu_s t^{2s}. \quad\text{.............(1)}$$

The quantity μ governs change of focus, ν_s governs spherical aberration of order $s-1$, so that (1) is the general expression required; and substituting directly in (6) § 3 we have the intensity given by

$$\sqrt{I}=\left|\int_0^1 e^{i\mu t}J_0(z\sqrt{t})\,dt\right|, \quad\text{....................(2)}$$

if we neglect aberrations. This expression therefore gives the intensity of illumination at any point of the neighbouring plane X; it will be seen that the co-ordinate ϕ_1 does not appear, so that the diffraction pattern is again a system of concentric circles. The general effect is shewn in the diagram and it is evident that there is a diminution of central intensity. If we adopt the Rayleigh limit, i.e., if we assume that a twenty per cent. loss in central intensity is allowable, we may find the corresponding change of focus permissible. But before this we may consider the extension of (2) to cover the presence of all orders of axial aberration; clearly it is

$$\sqrt{I}=\left|\int_0^1 e^{i\mu t+i\Sigma\nu_s t^{2s}}J_0(z\sqrt{t})\,dt\right|; \quad\text{.............(3)}$$

so that in this case also the diffraction pattern is a series of concentric circles.

Equation (2) may be evaluated by expansion in a series of Bessel functions and two series are obtained—one useful for large values of z

* Rayleigh, *Sci. Papers*, vol. III, p. 91.

and the other for small values*. These series are known as Lommel Functions—named after Lommel, who discussed them at some length.

The equation (3) may be evaluated in the same way and this gives rise to 'generalised Lommel Functions' which have been discussed in a paper by the author†; to consider them here would be beyond the limits of this tract.

7. *The Axial Intensity.* It is of importance to find the axial variation of intensity; we write $z=0$ in (3) § 6 and we have to consider the expression

$$\int_0^1 e^{i\mu t + i\Sigma\nu_s t^{2s}}\,dt,$$

where $\mu = 2\pi n$, $\nu_s = 2\pi n_s$ and n_s is the number of wave-lengths excess of aberration of order s, n denoting in the same way the change of focussing plane. If $\nu_s = 0$ we have

$$\int_0^1 e^{i\mu t}\,dt = \frac{e^{2\pi i n t}-1}{2\pi i n},$$

i.e.,

$$I = \left(\frac{\sin \pi n}{\pi n}\right)^2, \qquad \text{.........................(1)}$$

so that there are dark points along the axis given by integral values of n: moreover by writing $n=\frac{1}{4}$ we have $I = \cdot 81$ approximately, so that in accordance with the Rayleigh limit one-quarter of a wave-length change of focus is the utmost that can be allowed.

Taking into account now first order aberration we obtain the expression

$$\int_0^1 e^{i\mu t + i\nu_1 t^2}\,dt = \frac{1}{2\pi n}\int_0^{2\pi n} e^{i\left(v + \frac{n_1}{2\pi n^2}v^2\right)}\,dv;$$

so that by change of variable

$$\sqrt{I} = \left|\frac{1}{2\sqrt{n_1}}\int_{\frac{n}{\sqrt{n_1}}}^{\frac{n}{\sqrt{n_1}}+2\sqrt{n_1}} e^{\frac{i\pi t^2}{2}}\,dt\right|, \qquad \text{..............(2)}$$

and the value of this may be read off at once from tables of Fresnel's integrals. It follows immediately from this result, by writing $n+n_1=\pm\delta$, that the axial intensity is symmetrical about the point given by $n+n_1=0$ and this from § 4 is the mid-point of the longitudinal aberration; for the marginal focus is given by $n+2n_1=0$. This may be contrasted with the

* E. Lommel, *Abh. d. k. Bayer. Akad.* vol. xv, 1886. Cf. also Gray and Mathews, *Bessel Functions*, chap. xiv.

† 'Aberration Diffraction Effects,' *Phil. Trans. Roy. Soc.* A, 225.

result of the geometrical theory § 4, Chapter III, according to which the best focussing position is at a point of quadrisection of the longitudinal aberration and not as we see now at the centre.

If $n_1 = 1$ we have from (2) and tables of Fresnel's integrals the following:

$$n = \quad 0 \quad : \quad -\tfrac{1}{2} \quad : \quad -1 \quad : \quad -\tfrac{3}{2} \quad : \quad -2 \quad : \quad -3$$
$$I = 0{\cdot}0891 : 0{\cdot}3650 : 0{\cdot}8003 : 0{\cdot}3650 : 0{\cdot}0891 : 0{\cdot}0084$$

The intensity, therefore, at the paraxial focus falls to about 0·09 while if change of focus be admissible, to $n + 1 = 0$, the intensity rises to 0·80; and this illustrates the beneficial effect of change of focus. There are now no dark points upon the axis.

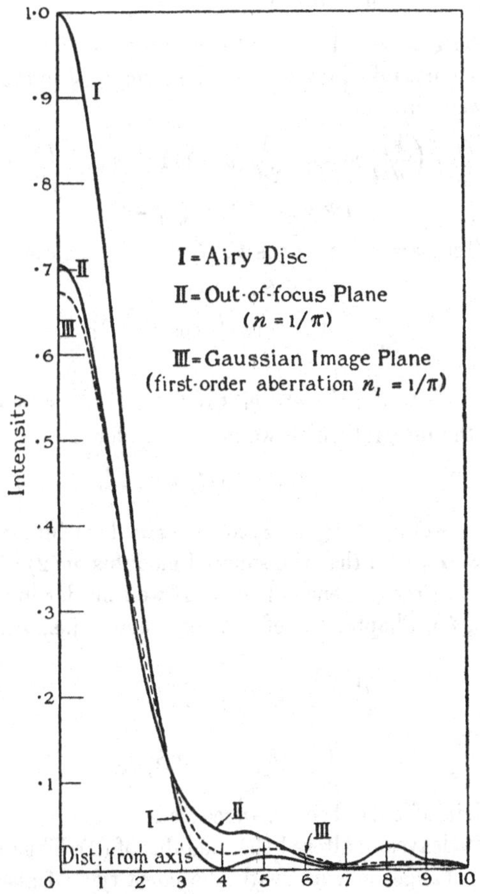

We may study in a similar way the effect of higher order aberrations; for taking account only of the first two orders it is usual to design a lens so that the paraxial and marginal foci coincide. It may be verified that with $2n_2 = 1$ the maximum intensity is about ·87 so that only a little more than half a wave-length of second order aberration can be admitted; but this is in the absence of first order aberration; in the presence of the latter as much as four wave-lengths of the second order aberration may be tolerated.

The diagram given shews the light distribution in the three cases marked; in each case a half-section is shewn passing through the centre of the system of contours; the general effect is seen to be to scatter the light into the outer parts of the field.

8. *Comatic Patterns.* In order to consider the effect of coma of the first order we must take into account the term in $\phi\psi$ in the function F so that in § 3 we write

$$\Psi = 2a_5\left(\frac{\rho}{d}\right)^3\left(\frac{Y_1}{d}\right)\cos\phi_1 + \frac{X}{2d^2}(\rho^2 - 2\rho Y_1 \cos\phi_1) - \frac{\rho\rho'}{d}\cos(\phi_1 - \phi'),$$

i.e.,
$$\kappa\Psi = \mu v + C\cos(\phi_1 - \psi), \quad \text{.....................(1)}$$

where
$$C\cos\psi = A - z\sqrt{v}\cos\phi', \quad C\sin\psi = -z\sqrt{v}\sin\phi',$$

and
$$A = \beta\sqrt{v^3} - \epsilon\sqrt{v},$$

so that
$$C^2 = A^2 - 2Az\sqrt{v}\cos\phi' + z^2 v.$$

Also

$$zd\sqrt{v} = \kappa\rho\rho', \; zd = \kappa\rho_1\rho', \; \beta d^4 = 2\kappa a_5 \rho_1^3 Y_1, \; \epsilon d^2 = \kappa X\rho_1 Y_1 \text{ and } 2\mu d^2 = \kappa X\rho_1^2.$$

In this case the integral which we must consider is

$$\int_0^1 e^{i\mu v} J_0(C)\,dv, \quad \text{..........................(2)}$$

where $C^2 = (\beta v - \epsilon)^2 v - 2zv(\beta v - \epsilon)\cos\phi' + z^2 v$. This integral reduces to unity if $\beta = z = \epsilon = 0$ so that the squared modulus of (2) will give us the light intensity in the presence of coma. If we consider in the first place the axis P_1C, § 5, Chapter III, of the coma figure, i.e., write $\phi' = 0$, we have

$$\int_0^1 e^{i\mu v} J_0(\beta v^{\frac{3}{2}} - z + \epsilon\sqrt{v})\,dv,$$

i.e.,
$$\int_0^1 e^{i\mu v} J_0(\beta v^{\frac{3}{2}} - z' v^{\frac{1}{2}})\,dv, \quad \text{.....................(3)}$$

where the origin of z has been changed; and it is clear that a change in the sign of μ leaves unaltered the modulus of (3). Thus nothing is to be gained by 'change of focus' with the coma-type of aberrations and this property is common to all the coma-type. We may therefore write $\mu = 0$ and consider only the effect upon the Gaussian plane—in which case (1) becomes

$$\sqrt{I} = \int_0^1 J_0(\sqrt{z^2 v - 2z\beta v^2 \cos\phi' + \beta^2 v^3})\, dv. \quad \text{............(4)}$$

It will be seen that $\beta d = \kappa\rho_1 P_1 G$, § 5, Chapter III, so that β is the measure of $P_1 G$ in the wave-length co-ordinates. The preceding integral may be evaluated by expansion in an infinite series of Bessel functions and in fact two series may be obtained suitable for various values of z, the one for the outer parts of the field and the other for the inner parts. Moreover the light contours may be drawn, i.e., the curves of constant intensity, by writing

$$\sqrt{I} = \text{const.}$$

in (4). But these expansions are beyond the scope of this tract*.

9. It is important to find the intensity upon the axis of the coma figure, i.e., when $\phi' = 0$; we have

$$\sqrt{I} = \int_0^1 J_0(\beta\sqrt{v^3} - z\sqrt{v})\, dv, \quad \text{.................(1)}$$

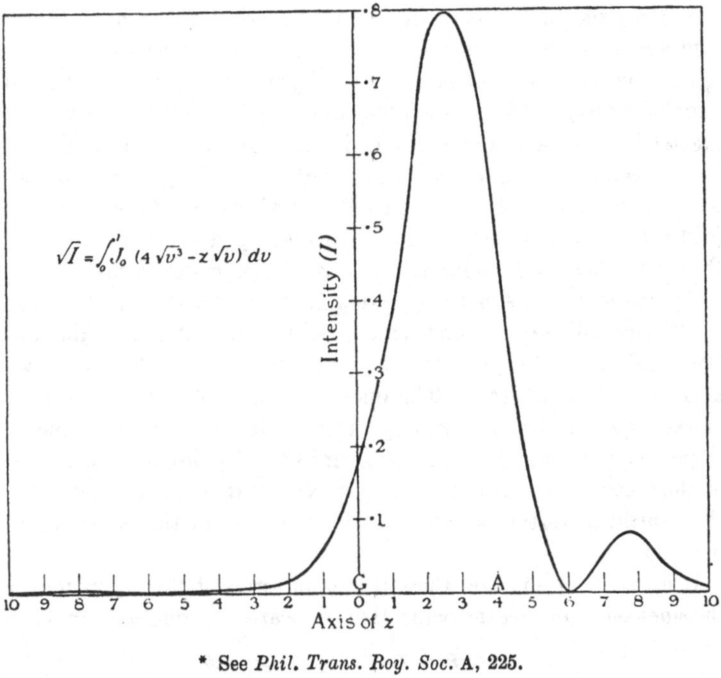

* See *Phil. Trans. Roy. Soc.* A, 225.

and here z may carry its own sign so as to cover the case of the line $\phi' = \pi$. Dark points occur therefore upon the axis of the figure, and their positions are given by the roots of the equation in z,

$$\int_0^1 J_0(\beta\sqrt{v^3} - z\sqrt{v})\, dv = 0; \quad \ldots\ldots\ldots\ldots\ldots\ldots(2)$$

and the positions of the turning values of the intensity are given by the roots of the equation in z,

$$\int_0^1 J_1(\beta v - z\sqrt{v^3})\, dv = 0; \quad \ldots\ldots\ldots\ldots\ldots\ldots(3)$$

and these in general will be the maxima of the intensity*.

It is of interest to examine some numerical results; thus let $\beta = 4$, then it will be found that the 'central' root of (3) is very nearly $z = 2\beta/3$ and that the intensity there is approximately ·80; the amount of coma therefore represented by $\beta = 4$ brings us to the Rayleigh limit. Moreover the central intensity has shifted from the conjugate point, towards the head of the coma figure; and as the amount of coma increases this central intensity point moves relatively towards the head of the figure; e.g., when $\beta = 12$ it is very nearly midway between P_1 and G.

If $\beta = 4$ the central contours are very nearly circular but the outer contours are elongated in the negative direction of the axis of the coma figure; and for larger values of β they depart very much more from the circular shape; moreover the intensities in one direction are very much greater than those in the opposite direction, and this produces the well-known 'comet' appearance associated with coma. The general conclusion may be drawn that there is no similarity whatever between the actual (diffraction) coma figure and that indicated by geometrical theory. On the other hand if z/β be large the figure approximates to the Airy disc.

Approximate roots of the equation (2) if $\beta = 4$ are $z = 11·4$, $7·7$, $2·7$, $-7·7$. The radius of the first light 'ring' is therefore 10·4 in the direction P_1G, and 5·0 in the opposite direction; while in the Airy disc it is 5·1 in either direction. The distortion effect of coma is thus very marked upon the first bright 'ring.' The radii of the first dark 'ring' are respectively 9·2 and 3·3, compared with the Airy disc 3·3; while those of the second 'ring' are 12 and 7 compared with 7. These results shew the central distortion which tends to disappear in the outer parts of the field.

It is useful to interpret these results in terms of the departure from the sine-condition: let the paraxial magnification produced by an optical

* See *Phil. Trans. Roy. Soc.* A, 225.

instrument be m and let the magnification calculated from the marginal ray be $m(1+\epsilon)$. Then we have

$$\epsilon Y_1 = \beta,$$

i.e., for the Rayleigh limit $\epsilon Y_1 = 4$. In the case therefore of a telescope objective $\epsilon = \frac{1}{300}$ would appear to give an ample field in which the Rayleigh limit is not exceeded. But of course in obtaining this we have taken into consideration only first order circular coma.

The diagram shews the light contours upon the Gaussian image plane in the presence of coma of amount given by $\beta = 4$.

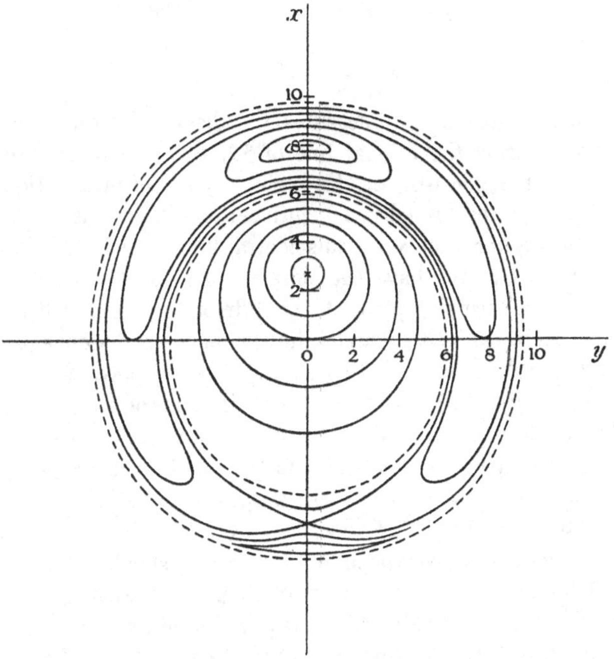

10. *Curvature of the Field and Astigmatism.* The coefficients governing curvature of the field and astigmatism in the function F are a_3 and a_4; so that from § 3 (4) we have

$$\Psi = \frac{X}{2d^2}(\rho^2 - 2\rho Y_1 \cos\phi_1) - \frac{\rho\rho'}{d}\cos(\phi_1 - \phi') + a_3\left(\frac{Y_1}{d}\right)^2\left(\frac{\rho}{d}\right)^2 + 4a_4\left(\frac{Y_1\rho}{d^2}\right)^2\cos^2\phi_1. \quad \text{......(1)}$$

If $a_4 = 0$ the two surfaces coincide and we are dealing simply with curvature of the field and the terms in (1) may be written

$$\frac{\rho^2}{2d^2}\left(X + 2a_3\frac{Y_1^2}{d^2}\right) - \frac{\rho C}{d}\cos(\phi_1 - \psi), \quad \text{..............(2)}$$

where $C\cos\psi = X\frac{Y_1}{d} + \rho'\cos\phi'$ and $C\sin\psi = \rho'\sin\phi'$. Now this is of the same form as § 6, so that we are led again to the same expression as that obtained for simple out-of-focus effects. Moreover if we write

$$Xd^2 + 2a_3 Y_1^2 = 0,$$

we have

$$\sqrt{I} = \int_0^{\rho_1} \rho\, d\rho \int_0^{2\pi} e^{-i\kappa\frac{\rho C}{d}\cos(\phi_1-\psi)}\, d\phi_1$$

$$= 2J_1\left(\frac{\kappa\rho_1 C}{d}\right)\Big/\frac{\kappa\rho_1 C}{d}, \quad \text{........................(3)}$$

i.e., the Airy disc; and C measures the distance of the point under consideration from the central line of § 2, Chapter III. The receiving plane in this position will pass through the point of intersection of the central line with the curved image surface: and in any case we have the law of intensity variation along this central line—and indeed in normal planes—as in § 7. We have therefore the general result; for varying positions of the receiving plane X the diffraction pattern will always be a ring system whose centre is upon the central line and for one particular plane this central intensity will be unity—i.e., we have the Airy disc. It is interesting to interpret these results in terms of the Petzval sum ϖ in order to discover what value of ϖ is permissible. As above, however, if one object only be present (a star in the case of a telescopic system), then a suitable change of focus will always give the Airy disc; but if several stars be in the field of view, to be examined simultaneously, then unlimited change of focus is not allowable. For one star disc will satisfy the Rayleigh limit for one set of positions of the plane X, another for another set, and so on, and that value of ϖ is required which will bring within the Rayleigh limit all the star discs for at least one position of the plane.

As in § 7 the wave-length change of focus allowable from the Airy disc is $n = \frac{1}{4}$; and since here $a_3 = \varpi d^2$ and $2FP = \varpi Y_1^2$ (§ 6, Chapter III) the limit will be satisfied provided

$$\frac{\kappa\rho_1^2}{2d^2}\cdot\frac{\varpi Y_1^2}{2} \not> \frac{\pi}{2},$$

i.e.,

$$\varpi Y_1^2 \not> \left(\frac{d}{\rho_1}\right)^2 \frac{2\pi}{\kappa},$$

i.e.,

$$\varpi Y_1^2 \not> \left(\frac{d}{\rho_1}\right)^2 \lambda, \quad \text{.............................(4)}$$

λ being the wave-length of the luminous vibration considered.

The focussing plane is movable, and this admits a value of ϖ twice as large as that indicated by (4); so that taking a numerical example in which $d = 30\rho_1$ and $\lambda/J = 0{\cdot}0006$ mm. if $Y_1/J = 2{\cdot}5$ cm. we have from (4), as the limiting value of ϖ,

$$J\varpi = {\cdot}0009,$$

where J is the modified power of the system. And this may be doubled as above.

11. In the general case, i.e., if $a_4 \neq 0$, astigmatism is present and we have (1) § 10: the resulting expressions admit of general integration but reference must be made to the paper quoted before*: some immediate specialisations of considerable importance however may be made. For example:

(1) the integrals are unaffected by a change in the sign of ϕ', so that the resulting light contours are symmetrical about the line $\phi' = 0$;

(2) they are unaffected by the addition of π to ψ; the origin is therefore a centre of symmetry of the contours;

(3) on differentiation with respect to ψ, and putting $\psi = 0$, it becomes evident that the contours cut the line $\psi = 0$ at right angles.

It is of importance to examine the intensity distribution along the central line. The general expression to be considered is

$$\mathcal{I} = \int_0^{\rho_1} \rho e^{i\gamma\rho^2} d\rho \int_0^{2\pi} e^{i\beta \sin^2 \phi_1} d\phi_1, \quad \text{..................(1)}$$

upon the central line, since here $C = 0$; where

$$2\gamma\rho^2 d^2 = \kappa\rho^2 \left\{X + \frac{2Y_1^2}{d^2}(a_3 + 4a_4)\right\} \text{ and } \beta d^4 = -4a_4\kappa\rho^2 Y_1^2.$$

Thus $$\gamma\rho^2 = \frac{\kappa\rho^2}{2d^2}\left\{X + \frac{2Y_1^2}{d^2}(a_3 + 2a_4)\right\} + 2a_4\frac{\kappa\rho^2 Y_1^2}{d^4} = \Gamma\rho^2 - \beta/2,$$

where Γ measures the separation of the focussing plane from the mid-point of the astigmatic separation. Then

$$\mathcal{I} = \int_0^{\rho_1} \rho e^{i\Gamma\rho^2} d\rho \int_0^{2\pi} e^{-i(\beta/2)\cos 2\phi_1} d\phi_1 = 2\pi \int_0^{\rho_1} \rho e^{i\Gamma\rho^2} J_0(a\rho^2)\, d\rho,$$

where $\beta = 2a\rho^2$; i.e.,

$$\mathcal{I} = \pi \int_0^{\rho_1^2} e^{i\Gamma u} J_0(au)\, du,$$

by change of variable. The intensity at the image point in the absence of aberrations must be unity, so that the expression whose squared modulus gives this intensity is

$$\frac{1}{\rho_1^2}\int_0^{\rho_1^2} e^{i\Gamma u} J_0(au)\, du. \quad \text{........................(2)}$$

* *Phil. Trans. Roy. Soc.* A, 225.

If now n be the number of wave-length units of astigmatism

$$2a\rho_1^2 = 2\pi n = 2\mu \text{ (say)},$$

and if n_1 measure the distance from the mid-point of the astigmatic separation

$$\Gamma\rho_1^2 = 2\pi n_1 = \lambda \text{ (say)},$$

and (2) becomes

$$\int_0^1 e^{i\lambda t} J_0(\mu t)\, dt, \quad \text{.........................(3)}$$

and this is the general expression whose squared modulus gives the intensity distribution along the central line—which evidently depends upon the *difference* between the curvatures of the two focal surfaces.

12. We consider some particular cases of (3) § 11.

(1) Let $\mu = 0$, i.e., let the system be free from astigmatism, but not necessarily from curvature of the field; we have

$$\int_0^1 e^{i\lambda t}\, dt,$$

leading at once to the case of (1) § 7.

(2) Let $\lambda = 0$, i.e., consider the intensity at the mid-point of the astigmatic separation; we have

$$\int_0^1 J_0(\mu t)\, dt.$$

Now $\int_0^1 J_0(1{\cdot}2t)\, dt = 0{\cdot}89$, so that in this case $I = 0{\cdot}79$: the quantity of astigmatism represented therefore by $\mu = 1{\cdot}2$ brings us to the Rayleigh limit for the mid-point. As a typical case let $d = 30\rho_1$ (in a telescope), then the astigmatic separation is given by

$$F_1F_2 = \frac{2\lambda}{2\pi}(30)^2 .\, 2{\cdot}4 = 0{\cdot}41 \text{ mm. approximately.}$$

(3) Let $\lambda \pm \mu = 0$, i.e., consider the intensity at either of the foci F_1 and F_2: we have

$$\int_0^1 e^{\mp i\mu t} J_0(\mu t)\, dt,$$

so that the intensities at the foci are equal; indeed, since in (3) § 11 a change in sign of λ leaves unaltered the modulus of the integral, the intensities are symmetrical about the mid-point of the astigmatic separation.

For the light contours and a more complete examination of the effects of astigmatism upon the diffraction pattern produced we must refer to the paper mentioned above.

13. *Distortion.* If all the aberration coefficients vanish except a_2 we are dealing with distortion, so that from (4) § 3 we have

$$\Psi = \frac{X}{2d^2}(\rho^2 - 2\rho Y_1 \cos\phi_1) - \frac{\rho\rho'}{d}\cos(\phi_1 - \phi') + 2a_2\left(\frac{Y_1}{d}\right)^2 \frac{Y_1 \rho}{d^2}\cos\phi_1, \quad \text{......(1)}$$

and if, in the first place, we write $X=0$ and consider only the paraxial image plane we have

$$\Psi = \frac{\rho C}{d}\cos(\phi_1 - \psi), \quad \text{......................(2)}$$

where $\quad C\cos\psi = \dfrac{2a_2}{d^3}Y_1^3 - \rho'\cos\phi', \quad C\sin\psi = -\rho'\sin\phi',$

so that $\quad C^2 = \left(\dfrac{2a_2}{d^3}Y_1^3\right)^2 - 2\rho'\cos\phi'\left(\dfrac{2a_2}{d^3}Y_1^3\right) + \rho'^2. \quad \text{.........(3)}$

From this last result it is evident that C measures the distance from the new (distortion) image point of § 7, Chapter III; and then from (2) we have

$$\sqrt{I} = \frac{1}{\pi\rho_1^2}\int_0^{\rho_1}\rho\, d\rho \int_0^{2\pi} e^{\frac{i\kappa\rho C}{d}\cos(\phi_1 - \psi)}\, d\phi_1,$$

i.e., $$\sqrt{I} = 2\frac{J_1\left(\dfrac{\kappa\rho_1 C}{d}\right)}{\dfrac{\kappa\rho_1 C}{d}}, \quad \text{......................(4)}$$

and accordingly we have the Airy disc with centre at the new distortion image point; the effect of distortion therefore is merely to move the ideal diffraction pattern as a whole through a distance equal and parallel to the geometrical distortion.

The general case for out-of-focus planes may be obtained directly from (1); for we have

$$\Psi = \frac{X}{2}\left(\frac{\rho}{d}\right)^2 + \frac{C'\rho}{d}\cos(\phi_1 - \psi'), \quad \text{.................(5)}$$

where C' is given by the relation (3) if for the terms in the brackets be substituted $\left(\dfrac{2a_2}{d^3}Y_1^3 - X\dfrac{Y_1}{d}\right)$. We are led again to the ordinary expression for out-of-focus effects—giving rise to Lommel functions. Moreover the centre of the diffraction pattern is upon the central line of the system; and the expression whose squared modulus gives the light intensity is

$$\frac{e^{2\pi i n}}{2\pi i n}\left\{(4\pi i n)\frac{J_1(z)}{z} - (4\pi i n)^2\frac{J_2(z)}{z^2} + \dots\right\}^*, \quad \text{.........(6)}$$

where $\kappa X\rho_1^2 = 4\pi n d^2$ and $\kappa\rho_1 C' = zd$.

* *Phil. Trans. Roy. Soc.* A, 225, p. 198.

REFERENCES

Airy. *Trans. Camb. Phil. Soc.* (1834), No. 283.
—— *Trans. Camb. Phil. Soc.* (1838).
—— *Phil. Mag.* 1841.
André. *Ann. d. l'École Normale Sup.* No. 5 (1876), No. 10 (1881).
Buxton, A. *Monthly Notices R. A. S.* LXXXI, 8 (1921).
—— *Monthly Notices R. A. S.* LXXXIII, 8 (1923).
Conrady, A. E. *Monthly Notices R. A. S.* LXXIX, 8 (1919).
Lommel. *Abh. d. k. Bayer. Akad.* XV (1886).
Martin, L. C. *Monthly Notices R. A. S.* LXXXII, 5 (1922).
Nagaoka. *Astrophysical Journal,* LI, 2.
Rayleigh. *Sci. Papers,* I, p. 415; III, p. 87 *et seq.*
Steward. *Phil. Trans. Roy. Soc.* A, 225 (1925).
—— *Proc. Opt. Conv.* (1926), I, p. 778.
Strehl. *Theorie des Fernrohrs* (Barth: Leipzig, 1897).
Struve. *Wied. Ann.* XVI (1882), p. 1008.
Walker. *Proc. Phys. Soc. Lond.* XXIV (1912), pp. 160–164.

CHAPTER VII

VARIOUS FORMS OF APERTURE

1. *The Annular Aperture.* In the preceding chapter we have everywhere assumed the stops of the system to be circular, and this will generally be the case; but occasionally other forms are used and in the present chapter accordingly is undertaken an examination of the diffraction patterns associated with these. As before we have the general expression (6) § 3, Chapter VI—the surface of integration being different. We consider three types in particular, viz.:

(1) the annular aperture—as suggested by Lord Rayleigh;

(2) a narrow rectangular aperture—as used in the determination of stellar diameters and for the separation of close double stars; and

(3) a semi-circular aperture—as in a heliometer.

For the first case let the central portion of the exit-pupil be blocked out so that the aperture consists of an annulus between circles of radii ρ_1 and $a\rho_1$, where $0 \leqslant a \leqslant 1$; then in the notation of the preceding chapter we have

$$S = \pi \rho_1^2 (1 - a^2),$$

giving the area of the exit-pupil; so that the general expression to be considered is

$$\sqrt{I} = \frac{2}{\rho_1^2(1-a^2)} \int_{a\rho_1}^{\rho_1} \rho J_0\left(\frac{\kappa\rho\rho'}{d}\right) d\rho = \frac{2}{1-a^2}\left\{\frac{J_1(z)}{z} - a^2 \frac{J_1(az)}{az}\right\}, \quad \ldots(1)$$

in the usual notation. This clearly reduces to the Airy disc when $a = 0$, and in any case the diffraction pattern consists of a series of concentric circles with centre at the Gaussian image. The radii of the dark rings are given by the roots of the equation in z,

$$J_1(z) - \alpha J_1(\alpha z) = 0, \quad\text{.........................(2)}$$

while the maxima are given by the roots of

$$J_2(z) - \alpha^2 J_2(\alpha z) = 0. \quad\text{.........................(3)}$$

The first root of (2) in the case $2\alpha = 1$ is just less than 3·15; with full aperture $\alpha = 0$ the corresponding root is 3·83. As $\alpha \to 1$ the expression (1) $\to J_0(z)$, the first root of which is 2·40; thus an increase in α gives a decrease in the radius of the first dark ring. Similarly for the first bright ring, for from (3) the first root is 4·80 if $2\alpha = 1$ and 3·83 if $\alpha = 1$; and these compare with the value 5·14 when $\alpha = 0$.

The annular aperture confers therefore a decided gain in resolving power, but the gain is counterbalanced by a loss of light due to the

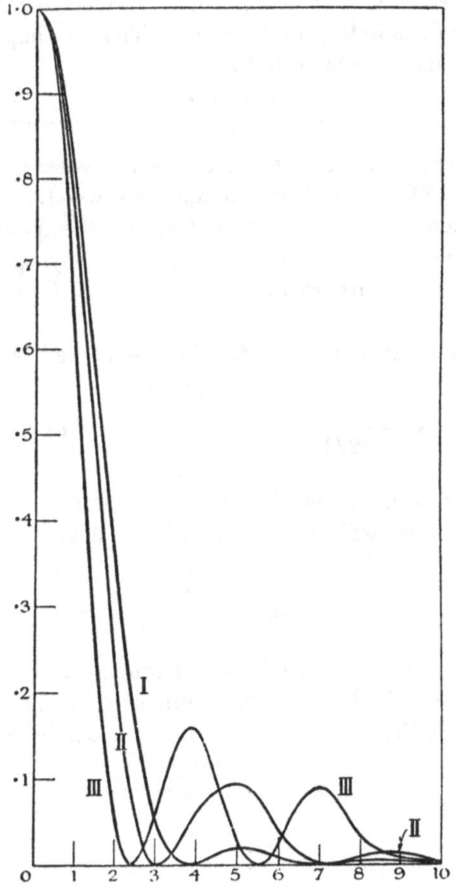

cutting down of the exit-pupil and thus the advantage obtainable depends upon the brightness of the object under examination.

The figure shews the distribution of intensity in the three cases I, $a=0$, II, $a=\frac{1}{2}$ and III, $a\to 1$; but it must be remembered that the relative intensity only is given, the central intensity being taken as unity in each case. To compare the curves one with another a factor $1-a^2$ must be applied.

2. If spherical aberration be present, combined with out-of-focus effects, the generalisation of (3) § 6, Chapter VI, is clearly

$$\frac{1}{1-a^2}\int_{a^2}^{1} e^{2\pi i(nt+n_1t^2)} J_0(z\sqrt{t})\,dt, \quad \text{.................}(1)$$

and the extension to cover the effect of axial aberration of any order is immediately obvious. Again the diffraction pattern is a ring system, as we should expect from symmetry; and to examine the distribution of axial intensity we have to write $z=0$ in (1). Thus for simple out-of-focus conditions the intensity is given by

$$\sqrt{I}=\frac{\sin\pi n(1-a^2)}{\pi n(1-a^2)}; \quad \text{.......................}(2)$$

and it is evident that the distance between successive dark points is given by $[n]=1/(1-a^2)$, and tends to infinity as $a\to 1$, as is evident also from the consideration that in the limiting case the aperture becomes a narrow rim. Again $n=1/4(1-a^2)$ brings us to the Rayleigh limit, so that the permissible departure from the Gaussian plane is increased in the ratio $1:1-a^2$.

Again if first order axial aberration be present we have

$$\sqrt{I}=\left|\frac{1}{2(1-a^2)\sqrt{n_1}}\int_{\frac{n}{\sqrt{n_1}}+2a^2\sqrt{n_1}}^{\frac{n}{\sqrt{n_1}}+2\sqrt{n_1}} e^{\frac{i\pi t^2}{2}}\,dt\right|, \quad \text{...........}(3)$$

and the value of this may be obtained at once from tables of Fresnel's integrals. As a numerical example we find that if $2a^2=1$, $n+1=0$ and $n_1=1$ in (3),

$$\sqrt{I}=\left|\int_0^1 e^{\frac{i\pi t^2}{2}}\,dt\right|,$$

so that $I=\cdot 80$; thus one wave-length of primary aberration combined with one wave-length change of focus brings us to the limit. It is of interest to notice that at the Gaussian image in this case we have

$$\sqrt{I}=\left|\int_1^2 e^{\frac{i\pi t^2}{2}}\,dt\right|,$$

so that $I = \cdot 09$; here again, therefore, the beneficial effects of a slight change of focus are manifest.

3. If coma only be present we have seen that the most important case to consider is that of the intensity distribution upon the Gaussian plane and the generalisation for the annular aperture is clearly

$$\sqrt{I} = \frac{1}{1-a^2}\int_{a^2}^{1} J_0(\sqrt{z^2 v - 2z\beta v^2 \cos\phi' + \beta^2 v^3})\,dv; \quad \text{......(1)}$$

so that the intensity upon the axis of the coma figure is given by writing $\phi' = 0$ in (1); and the dark points and turning values of I are given respectively by the equations in z,

$$\int_{a^2}^{1} J_0(\beta\sqrt{v^3} - z\sqrt{v})\,dv = 0, \quad \text{......(2)}$$

and

$$\int_{a^2}^{1} J_1(\beta v - z v^{\frac{1}{3}})\,dv = 0. \quad \text{......(3)}$$

As before the latter equation will give maxima of I in general.

We know that in the case of the full circular aperture $\beta = 4$ brings us to the Rayleigh limit; but from (1) if $2a^2 = 1$, $\beta = 4$, $z = 3$ we have

$$\sqrt{I} = 2\int_{\cdot 5}^{1} J_0(4\sqrt{v^3} - 3\sqrt{v})\,dv = \cdot 95,$$

so that $I = \cdot 91$ and larger values of β are admissible. Indeed putting $\beta = 6$, $2a^2 = 1$ and $z = 4 \cdot 75$ we have $I = \cdot 76$; so that $\beta = 6$ brings us approximately

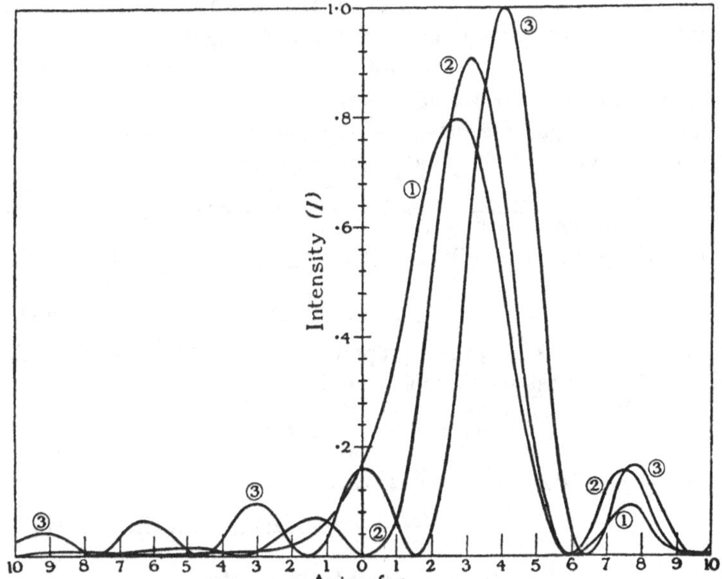

to the limit in this case. The annular aperture leads to a smaller relative loss of intensity for a given field, or, on the other hand, it increases the field of view subject to the Rayleigh limit. In the limiting case when the aperture becomes a narrow rim we have

$$\sqrt{I} \to J_0(\beta - z); \quad \text{.........................(4)}$$

but the absolute intensity tends to zero. This result indicates a relative maximum of unity for all values of β, i.e., for all amounts of coma present. (4) refers only to the axis of the coma figure; more generally we have

$$\sqrt{I} \to J_0(u),$$

where $u^2 = z^2 - 2z\beta \cos \phi' + \beta^2$ and so u is the distance from the point $(\beta, 0)$ in the coma figure. The diffraction pattern therefore tends to become a series of concentric circles with this point as centre and the radii of the first bright and dark rings are respectively 2·405 and 3·832: this illustrates the decided gain in resolving power over the full aperture case.

These results are illustrated in the figure, which shews also the change in the position of the central maximum.

4. The effect of curvature of the field can be deduced from the preceding paragraphs, for it has been seen that, with full aperture, curvature gives ordinary out-of-focus effects combined with a change of the centre of the ring system. A numerical value of ϖ therefore is admissible with the annular aperture greater than that with the full aperture in the ratio $1 : 1 - a^2$. Thus in the case considered in the last chapter a sufficiently large field is given by the limit

$$(1 - a^2) J\varpi \not> 0{\cdot}0018,$$

J being the modified power of the system.

As regards astigmatism the extension of the foregoing results is evident; the general expression for points upon the central line is*

$$\frac{1}{\pi \rho_1^2 (1 - a^2)} \int_{a\rho_1}^{\rho_1} \rho e^{i\Gamma\rho^2} d\rho \int_0^{2\pi} e^{-i(\beta/2)\cos 2\phi_1} d\phi_1, \quad \text{.........(1)}$$

so that the intensity distribution along this line is given by

$$\sqrt{I} = \frac{1}{1 - a^2} \int_{a^2}^{1} e^{i\lambda t} J_0(\mu t)\, dt, \quad \text{.....................(2)}$$

and our remarks as to symmetry apply as before. The intensity at the mid-point of the astigmatic separation is

$$\sqrt{I} = \frac{1}{1 - a^2} \int_{a^2}^{1} J_0(\mu t)\, dt, \quad \text{.......................(3)}$$

while the intensity at either focus is

$$\sqrt{I} = \frac{1}{1 - a^2} \int_{a^2}^{1} e^{\pm i\mu t} J_0(\mu t)\, dt. \quad \text{..................(4)}$$

* In the notation of §§ 11, 12, Chapter VI.

As a numerical example let us write in (3), $2a^2=1$, $\mu=1{\cdot}2$—this value of μ leads us to the Rayleigh limit with the full circular aperture—then $I=\cdot 64$: thus here too large a value of μ has been taken : the annular aperture is seen to be unfavourable in the case of astigmatism—at least as far as the central intensity is concerned. This result might have been anticipated from the form of (3); for as $a \to 1$ we see that $\sqrt{I} \to J_0(\mu)$ and the value $\mu=\cdot 661$ gives the Rayleigh limit here: and for any other value of a the limiting value of μ will be between $1{\cdot}2$ and $\cdot 661$.

Distortion need not be considered in detail for its effect is but to move the 'ideal' diffraction pattern as a whole.

5. Hitherto, we have considered only the diffraction effects of the separate aberrations and it is evident that their combination will lead to complicated expressions; but if the exit-pupil be a narrow rim the combined effects may be considered more simply. For with the usual notation the expression to be considered is

$$\frac{1}{2\pi}\int_0^{2\pi} e^{i\kappa\Psi}\,d\phi_1. \qquad \text{.........................(1)}$$

From § 3, Chapter VI, we have

$$\kappa\Psi = -\zeta\cos(\phi_1-\psi) - 2\mu\cos^2\phi_1,$$

where ζ involves the coefficients a_2 and a_5 and μ the coefficient a_4, i.e., astigmatism; and ζ, ψ are polar co-ordinates referred to a new origin in the receiving plane. The expression (1) becomes now

$$\frac{1}{2\pi}\int_0^{2\pi} e^{-i\zeta\cos(\phi_1-\psi)-2i\mu\cos^2\phi_1}\,d\phi_1, \qquad \text{..............(2)}$$

and if $\mu=0$, i.e., if the system be free from astigmatism,

$$\sqrt{I}=J_0(\zeta), \qquad \text{..........................(3)}$$

and $J_0(\zeta) \to \sqrt{\dfrac{2}{\pi\zeta}}\cos(\zeta-\tfrac{1}{4}\pi)$ when ζ is large, so that the mean brightness varies as $1/\zeta$.

The effect therefore of the primary aberrations, apart from astigmatism, is merely to move the diffraction pattern bodily through a distance depending upon the amount of aberration present. We may contrast this simple effect with that found in § 15, Chapter III, for the aberration curves corresponding to circular coma, on a purely geometrical ray theory. The general evaluation of (2) may be obtained as indicated in the preceding chapter; but we can obtain the information we desire by writing $\psi = 0, \pi/2$: then

$$\frac{1}{2\pi}\int_0^{2\pi} e^{-i\zeta\sin\phi_1 - 2i\mu\cos^2\phi_1}\,d\phi_1 = \frac{1}{\pi}\sum_{n=0}(-2i\mu)^n \frac{\lfloor 2n}{2^n(\lfloor n)^2}\frac{J_n(\zeta)}{\zeta^n} \quad \text{...(4)}$$

if $\psi = \pi/2$; while if $\psi = 0$ we have

$$\frac{1}{2\pi}\int_0^{2\pi} e^{-i\zeta\cos\phi_1 + 2i\mu\sin^2\phi_1}\,d\phi_1 = \sum_{n=0}^{\infty}(2i\mu)^n \frac{\lfloor 2n}{2^n(\lfloor n)^2}\frac{J_n(\zeta)}{\zeta^n} \quad \text{.......(5)}$$

6. *The Slit Aperture.* Let the exit-pupil consist of two parallel rectangular apertures symmetrically placed with respect to the axis of

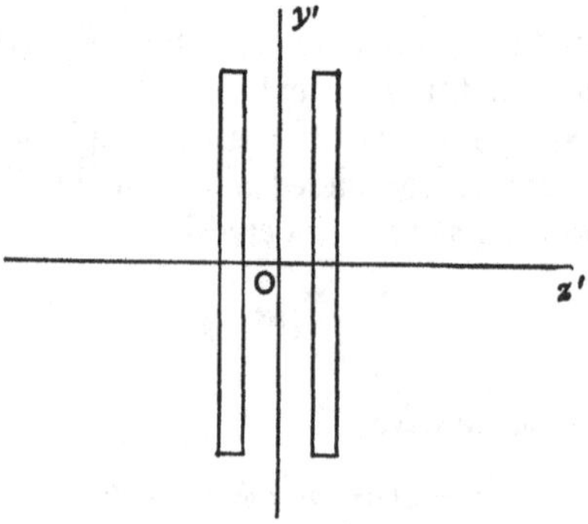

y' and extending from $y' = +a$ to $y' = -a$; let the breadth be given by $z = b, b'$ and $z = -b, -b'$, so that $S = 4a(b - b')$. Thus in the absence of aberrations ($F = 0$) and upon the paraxial image plane ($X = 0$) the integral which we have to consider is

$$\frac{1}{4a(b-b')}\iint e^{-i\kappa(\rho\rho'/d)\cos(\phi_1 - \phi')}\,dy'dz' = \frac{1}{4a(b-b')}\iint e^{-i(\kappa/d)(hy'+kz')}\,dy'dz' \quad \text{......(1)}$$

if h, k be the co-ordinates of the pole at which the light intensity is required; and (1) takes the form

$$\frac{\sin(\kappa ah/d)}{(\kappa ah/d)}\left\{\frac{\sin(\kappa bk/d)}{\kappa k/d} - \frac{\sin(\kappa b'k/d)}{\kappa k/d}\right\}\frac{1}{b-b'} \quad \text{.........(2)}$$

The case of practical importance arises when $k = 0$, i.e., we consider the intensity upon a line in the receiving plane parallel to the y' axis above: and then

$$\sqrt{I} = \frac{\sin(\kappa ah/d)}{\kappa ah/d} = \frac{\sin z}{z}, \quad \text{..................(3)}$$

where z is the usual wave-length co-ordinate. It is evident that an expression of precisely the same form arises if we consider only one symmetrical slit instead of the two above.

From (3) the intensity vanishes at the points given by $z = n\pi$, where n is any positive or negative integer other than zero; and the maxima occur at the points given by the roots of the equation in z,

$$\tan z = z. \quad \text{.................................(4)}$$

It will be seen that these two equations replace those of the preceding chapter

$$J_1(z) = 0 \text{ and } J_2(z) = 0. \quad \text{.....................(5)}$$

Comparing the minima we have from the above

3·142, 6·283 and 9·425,

while for the full circular aperture the corresponding values are

3·832, 7·016 and 10·173:

for the maxima they are

4·494, 7·725 and 10·943,

and 5·135, 8·417 and 11·620,

respectively.

The rectangular aperture, therefore, confers a very distinct advantage in the resolution of close double stars, provided always that the loss of light be not an insuperable difficulty.

7. In the case of simple out-of-focus conditions a term $\kappa X\rho^2/2d^2$ must be added to the index of the integrand, or, since we integrate here with respect to y' only, a term of the form

$$\kappa X y'^2/2d^2.$$

The appropriate expression is therefore

$$\frac{1}{2a}\int e^{\frac{i\kappa Xy'^2}{2d^2} - \frac{i\kappa\rho\rho'}{d}\cos(\phi_1 - \phi')} dy', \quad \text{..................(1)}$$

and the light intensity is given by the squared modulus of the expression

$$\int_0^1 e^{i(\mu v^2 - zv)}\, dv\text{*}, \quad \text{..........................(2)}$$

where μ denotes the displacement of the receiving plane. Putting $z = 0$ we have

$$\int_0^1 e^{i\mu v^2} dv,$$

and the axial distribution of intensity is given by

$$\sqrt{I} = \left| \frac{1}{2\sqrt{n}} \int_0^{2\sqrt{n}} e^{\frac{i\pi v^2}{2}}\, dv \right|, \quad \text{....................(3)}$$

* We assume that $\phi' = 0$, so that the intensity is considered only for points along a line parallel to the slit aperture.

where $\mu = 2\pi n$; and the value of this may be read off at once from tables of Fresnel's integrals. In general if $z = -2\pi N$ we have

$$\sqrt{I} = \left| \frac{1}{2\sqrt{n}} \int_{\frac{N}{\sqrt{n}}}^{2\sqrt{n}+\frac{N}{\sqrt{n}}} e^{\frac{i\pi t^2}{2}} dt \right|, \quad \text{.................(4)}$$

by means of the substitution $v\sqrt{\mu} - z/2\sqrt{\mu} = t\sqrt{\pi/2}$; and this also is readily obtainable from tables of Fresnel's integrals.

The extension of these results to take into account the presence of spherical aberration leads clearly to the expression

$$\int_0^1 e^{i(-zv+\mu v^2+\nu v^4+\dots)} dv, \quad \text{.......................(5)}$$

where ν, ... denote aberrations of higher orders. In the particular case of first order aberration only we have

$$\int_0^1 e^{i(\mu v^2+\nu v^4)} dv = \int_0^1 \sum_{n=0}^{\infty} \frac{y_n}{\lfloor n} v^{2n} dv = \sum_{n=0}^{\infty} \frac{y_n}{\lfloor n} \frac{1}{2n+1}, \quad \text{.........(6)}$$

upon the axis of the system; where

$$y_n = (i\mu)^n + \frac{n^2}{2}(i\mu)^{n-2}\frac{i\nu}{2} + \dots + \frac{n^{2r}}{2.4\dots 2r}(i\mu)^{n-2r}\left(\frac{i\nu}{2}\right) + \dots.$$

From (3) if $n = \frac{1}{4}$ we have

$$I = \left| \int_0^1 e^{i\pi v^2/2} dv \right|^2 = \cdot 8319,$$

so that $n = \frac{1}{4}$ brings us practically to the Rayleigh limit; this then is the maximum shift of focus plane permissible. Again for the point $z\,(=-2\pi N)$ upon the plane we have

$$I = \left| \int_{2N}^{1+2N} e^{i\pi v^2/2} dv \right|^2,$$

and from tables the following values appear:

$N=$	0	·1	·2	·3	·4	·5	·6	·7	·8
$I=$	·8319	·6486	·4837	·3251	·1902	·0883	·0344	·0089	·0207

The first maximum is given, therefore, by $N = \cdot 73$, approximately, and its radius is $2\pi \times \cdot 73$, i.e., 4·6.

The general axial intensity in the presence of first order aberration given by $\nu = 2\pi n_1$ is given by the squared modulus of

$$\int_0^1 e^{2\pi i(nv^2+n_1 v^4)} dv.$$

Here $n_1 = \frac{7}{4}$ and $n = -\frac{7}{4}$ gives very nearly the Rayleigh limit, so that with change of focus one-and-three-quarter wave-lengths of first order spherical aberration are permissible; it will be remembered that with the full circular aperture one wave-length only of first order aberration is permissible.

8. If coma alone be present we have from Chapter VI, § 3

$$\Psi = 2a_5 \frac{Y_1 \rho^3}{d^4} \cos \phi_1 - \frac{\rho \rho'}{d} \cos \phi_1,$$

where we have written $\phi' = 0$;

i.e.,
$$\Psi = 2a_5 \frac{Y_1}{d^4} y' (y'^2 + z'^2) - \frac{\rho' y'}{d},$$

if we use the variables appropriate to the slit aperture. Neglecting the z' dimension we are led to an expression of the form

$$\int_{-a}^{+a} e^{i(Ay^3 - By)}\, dy, \quad \text{.........................(1)}$$

where A and B are constants; and since the function in the index of the exponential is odd this gives—apart from a numerical factor—

$$\int_0^a \cos (Ay^3 - By)\, dy, \quad \text{.......................(2)}$$

so that in the usual notation the light intensity at a point z in the presence of an amount of coma represented by β is given by

$$\sqrt{I} = \int_0^1 \cos (\beta u^3 - zu)\, du. \quad \text{....................(3)}$$

This result applies only to the Gaussian image plane but change of focus may be included by the addition of a term μu^2 to the argument of the circular function; but as indicated in § 8, Chapter VI, the effect upon the Gaussian plane is the most important. The positions of the turning values of I, and these in general will be maxima, are given by the roots of the equation in z

$$\int_0^1 \sin (\beta\sqrt{u^3} - z\sqrt{u})\, du = 0. \quad \text{....................(4)}$$

From (3) if $\beta = 3{\cdot}25$ the maximum value of the intensity occurs at the point $z = 1{\cdot}75$ and it is ·78 times the intensity at the theoretical image in the absence of coma; this degree of coma therefore brings us to the limit for the slit aperture. $\beta = 4$ brought us to the limit with the full circular aperture so that to this extent the slit aperture is seen to be unfavourable with coma.

But the radius of the first dark 'ring' in the direction away from the head of the geometrical coma figure is 2·75 here, as compared with 3·35 with the full circular aperture; in the Airy disc this same radius is 3·83. For the first bright ring the figures are respectively 4·75; 5·35; 5·14.

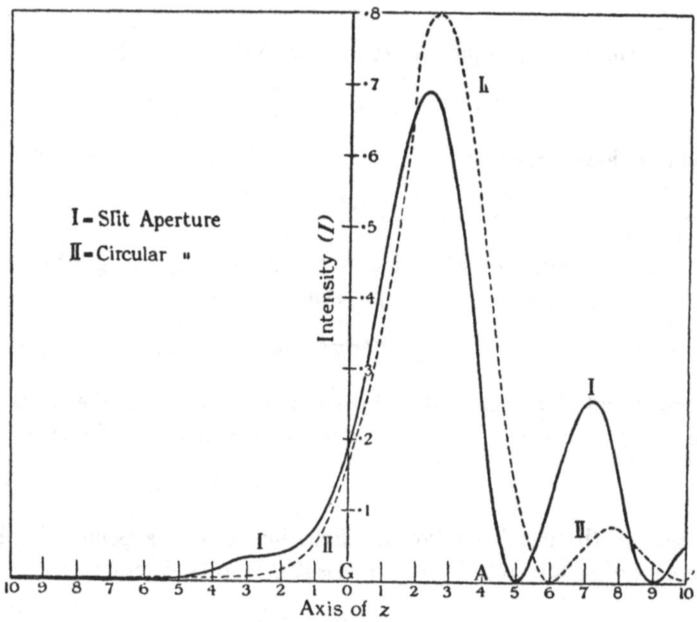

The diagram shews the intensity distribution along the line in the image plane parallel to the aperture, in the case $\beta = 4$; and the dotted curve shews the distribution along the same line when the aperture is a circle and coma is present to the same degree.

9. With curvature and astigmatism we have

$$\Psi = \frac{X}{2d^2}(\rho^2 - 2\rho Y_1) - \frac{\rho\rho'}{d} + (a_3 + 4a_4)\left(\frac{Y_1\rho}{d}\right)^2,$$

where we have written $\phi' = 0$ and $\phi_1 = 0$; so that upon the central line given by $\rho' d + XY_1 = 0$, we have an expression of the form

$$\int_{-a}^{+a} e^{iA\rho^2}\, d\rho,$$

reducing in the usual way to

$$\int_0^1 e^{i\lambda u^2}\, du, \qquad \text{..............................(1)}$$

where λ as usual measures the distance from the primary focus. If the slit aperture be rotated through a right angle the expression is

$$\int_0^1 e^{i\lambda' u^2} du, \quad \text{.............................(2)}$$

where λ' measures the distance from the secondary focus. Away from the central line we have, corresponding to (1), the integral

$$\int_0^1 e^{i\lambda u^2 - izu} du. \quad \text{.............................(3)}$$

This is of the form discussed in the preceding chapter; and it may be evaluated immediately from tables of Fresnel's Integrals.

It is evident that the function of the slit aperture is to eliminate rays travelling to the focal line parallel to it; simple out-of-focus conditions are therefore reproduced.

The effect of distortion is merely to move the diffraction pattern as a whole and the results of § 7 may therefore be quoted.

10. *The Semi-Circular Aperture.* Let the exit-pupil be semi-circular in form; then two cases arise: we may assume the Gaussian image point to lie upon the projection in the receiving plane of the diameter of the semi-circle, in which case $0 \leqslant \phi_1 \leqslant \pi$, in the usual notation; or we may suppose the image point to lie upon a perpendicular line, i.e., $-\pi/2 \leqslant \phi_1 \leqslant \pi/2$. Intermediate cases occur but the foregoing are the only two which we shall consider. And taking the first possibility we are led, for an aberrationless system, to the expression

$$\int_0^{\rho_1} \rho d\rho \int_0^{\pi} e^{-it\cos(\phi_1 - \phi')} d\phi_1, \quad \text{..................(1)}$$

where $td = \kappa\rho\rho'$.

Now $$\cos(x\cos\theta) = J_0(x) + 2\sum_{n=1}^{\infty}(-1)^n J_{2n}(x)\cos 2n\theta,$$

and $$\sin(x\cos\theta) = 2\sum_{n=0}^{\infty}(-1)^n J_{2n+1}(x)\cos(2n+1)\theta;$$

so that

$$\int_0^{\pi} e^{-it\cos(\phi_1 - \phi')}dt = \int_0^{\pi}\cos(t\cos\overline{\phi_1 - \phi'})\,d\phi_1 - i\int_0^{\pi}\sin(t\cos\overline{\phi_1 - \phi'})\,d\phi_1$$

$$= \pi J_0(t) - 4i\eta(t), \quad \text{..................................(2)}$$

where $$\eta(t) = J_1(t)\sin\phi' - \frac{J_3(t)\sin 3\phi'}{3} + \frac{J_5(t)\sin 5\phi'}{5} - \dots \quad (3)$$

If we write, as usual, $zd = \kappa\rho_1\rho'$ then the expression whose squared modulus gives the light intensity is

$$\frac{2}{\pi z^2}\int_0^z \{\pi J_0(t) - 4i\eta(t)\}\, t\,dt. \quad \text{......................(4)}$$

It is seen that this reduces to unity at the Gaussian image, where $z = 0$; we have, therefore,

$$\frac{2J_1(z)}{z} - \frac{8i}{\pi z^2}\int_0^z \eta(t)\, t\,dt,$$

i.e.,

$$\frac{2J_1(z)}{z} - \frac{8i}{\pi z^2}\zeta(z), \quad \text{..........................(5)}$$

if

$$\zeta(z) = \int_0^z \eta(t)\, t\,dt.$$

Thus

$$I = \left\{\frac{2J_1(z)}{z}\right\}^2 + \frac{64}{\pi^2}\left\{\frac{\zeta(z)}{z^2}\right\}^2. \quad \text{..................(6)}$$

11. If $\phi' = 0$, i.e., if the pole lie upon the edge of the projection of the aperture upon the receiving plane, then $\eta(t) = 0$; so that (6) reduces to

$$I = \left\{\frac{2J_1(z)}{z}\right\}^2,$$

and this is as in the Airy disc. If $\phi' = \pi/2$, then

$$\eta(t) = J_1(t) + \tfrac{1}{3}J_3(t) + \tfrac{1}{5}J_5(t) + \dots,$$

so that

$$\zeta(z) = \int_0^z \eta(t)\, t\,dt = z\sum_{n=1}^{\infty}\frac{J_{2n}(z)}{2n-1} + \tfrac{1}{2}\int_0^z \{1 - J_0(t)\}\,dt, \quad \text{......(1)}$$

since

$$\int_0^z \frac{J_{2n-1}(t)}{2n-1}\, t\,dt = \frac{zJ_{2n}(z)}{2n-1} + \int_0^z J_{2n}(t)\,dt,$$

and

$$1 - J_0(z) = 2\sum_{n=1}^{\infty} J_{2n}(z).$$

Thus (1) combined with (6) of the preceding paragraph gives the light intensity at distance z from the Gaussian image upon the line $\phi' = \frac{\pi}{2}$*.

It is worthy of notice that the maxima and minima of the intensity are now at the points given by the roots of the equation in z

$$\frac{\pi^2}{16}\frac{J_1(z)J_2(z)}{\zeta(z)} = \frac{d}{dz}\left\{\frac{\zeta(z)}{z^2}\right\}. \quad \text{......................(2)}$$

From the preceding it is evident that the effect of the aperture is to scatter light into the outer parts of the field. There will be no dark points now upon the line $\phi' = \pi/2$, nor, indeed, anywhere except upon the line $\phi' = 0$ (or π).

* Formulae equivalent to these were given by Bruns, *Astron. Nachr.* vol. CIV, p. 1 (1883).

12. In the presence of coma, we have, upon the paraxial image plane,

$$\Psi = -\frac{C\rho}{d}\cos(\phi_1 - \psi),$$

where $\quad C\cos\psi = \rho'\cos\phi' - 2a_s\gamma_1\rho^2/d^3, \quad C\sin\psi = \rho'\sin\phi'$
from § 3, Chapter VI; so that we have to consider the expression

$$\frac{2}{\pi\rho_1^2}\int_0^{\rho_1}\rho\,d\rho\int_0^{\pi} e^{-i\frac{\kappa C\rho}{d}\cos(\phi_1-\psi)}\,d\phi_1,$$

i.e.,
$$\frac{2}{\pi\rho_1^2}\int_0^{\rho_1}\rho\,d\rho\left\{\pi J_0\left(\frac{\kappa C\rho}{d}\right) - 4i\eta\left(\frac{\kappa C\rho}{d}\right)\right\}*$$

from (2) § 10.

By means of the substitution $\kappa\rho\rho' = zd\sqrt{v}$, $zd = \kappa\rho_1\rho'$ this reduces to

$$\int_0^1 J_0(u)\,dv - \frac{4i}{\pi}\int_0^1 \eta(u)\,dv, \quad \text{......................(1)}$$

where $u^2 = z^2v - 2z\beta v^2\cos\phi' + \beta^2 v^3$ and β, as usual, is the measure of the coma present; the light intensity is given by the squared modulus of (1). Upon the axis of the coma figure $\phi' = 0$, and then $\eta(u) = 0$; so that

$$\sqrt{I} = \int_0^1 J_0(z\sqrt{v} - \beta\sqrt{v^3})\,dv, \quad \text{......................(2)}$$

and this is the same formula as with the full circular aperture. At right-angles to the axis of the coma figure $\phi' = \pi/2$, and (1) becomes

$$\int_0^1 J_0(u)\,dv - \frac{4i}{\pi}\int_0^1 \eta(u)\,dv, \quad \text{......................(3)}$$

where $u^2 = z^2v + \beta^2v^3$.

Here the effect of closing down the aperture is seen in the presence of the second integral.

13. Let us now take the second possibility of § 10, i.e., let us assume that $-\pi/2 \leqslant \phi_1 \leqslant \pi/2$; then we are led to the formula

$$\frac{2}{\pi\rho_1^2}\int_0^{\rho_1}\rho\,d\rho\int_{-\pi/2}^{\pi/2} e^{-i\frac{\kappa C\rho}{d}\cos(\phi_1-\psi)}\,d\phi_1. \quad \text{...............(1)}$$

Now
$$\int_{-\pi/2}^{\pi/2} e^{-it\cos(\phi_1-\psi)}\,d\phi_1 = \pi J_0(t) - 4i\eta_1(t),$$

where

$$\eta_1(t) = J_1(t)\cos\psi + \tfrac{1}{3}J_3(t)\cos 3\psi + \tfrac{1}{5}J_5(t)\cos 5\psi + \dots.$$

* The expression for $\eta\left(\frac{\kappa C\rho}{d}\right)$ contains ψ instead of ϕ', as in § 10, (3).

The expression, therefore, whose squared modulus gives the light intensity is

$$\int_0^1 J_0(u)\,du - \frac{4i}{\pi}\int_0^1 \eta_1(u)\,du,$$

where u has the meaning of the previous paragraph. If $\phi' = 0$ then $\psi = 0$ and

$$\eta_1(t) = J_1(t) + \frac{J_3(t)}{3} + \frac{J_5(t)}{5} + \dots,$$

and we have

$$\int_0^1 J_0(z\sqrt{v} - \beta\sqrt{v^3})\,dv - \frac{4i}{\pi}\int_0^1 \eta_1(z\sqrt{v} - \beta\sqrt{v^3})\,dv; \quad \text{.........(2)}$$

while if $\phi' = \pi/2$ we obtain

$$\int_0^1 J_0(\sqrt{z^2 v + \beta^2 v^3})\,dv - \frac{4i}{\pi}\int \eta_1(\sqrt{z^2 v + \beta^2 v^3})\,dv. \quad \text{.........(3)}$$

14. As we have seen curvature of the field and astigmatism are governed by the coefficients a_3 and a_4; with curvature alone we have simple out-of-focus effects and the preceding results may be quoted. But with the introduction of astigmatism a new integral presents itself. As in § 11 of Chapter VI we have the expression

$$\int_0^{\rho_1} \rho e^{i\gamma\rho^2} d\rho \int_{-\pi/2}^{\pi/2} e^{-i\zeta\cos(\phi_1 - \psi) + i\beta\sin^2\phi_1}\,d\phi_1,$$

where now ζ, ψ are the polar co-ordinates of the point at which the intensity is to be estimated—the origin being taken upon the central line; and the range of integration for ϕ_1 is from $-\pi/2$ to $\pi/2$. If $\zeta = 0$, i.e., upon the central line the integral may be evaluated as previously; but in the general case, $\zeta \neq 0$, we are led to an expression in Struve's Functions and reference must be made to the above-mentioned paper*.

Again, the effect of distortion is merely to move bodily the whole diffraction pattern produced by an aberrationless system and therefore no further investigation is required.

REFERENCES

Rayleigh. *Sci. Papers*, III, p. 90.
Steward. *Phil. Trans.* A, 225 (1925).
—— *Proc. Opt. Convention* (1926).

* *Phil. Trans.* A, 225 (1925), Part III.

Printed in the United States
By Bookmasters